Introduction
to
Forensic Engineering

Randall Noon
Dressler Consulting Engineers, Inc.
Overland Park, Kansas

CRC Press
Boca Raton   Ann Arbor   London   Tokyo

Library of Congress Cataloging-in-Publication Data

Noon, Randall.
Introduction to forensic engineering / Randall Noon.
     p. cm.
  Includes bibliographical references and index.
  ISBN 0-8493-8102-9
  1. Forensic engineering. I. Title.
TA219.N66   1992
620—dc20   92-19405
                                      CIP

© 1992 by CRC Press, Inc.

International Standard Book Number 0-8493-8102-9

Library of Congress Card Number 92-19405
Printed in the United States   3 4 5 6 7 8 9 0
Printed on acid-free paper

THE FORENSIC LIBRARY
Medicine, Law, Sciences

Editor-in-Chief
William G. Eckert

Managing Editor
George Novotny

Introduction to Forensic Engineering
Randall Noon
Dressler Consulting Engineers, Inc.

Introduction to Forensic Sciences, 2nd Edition
William G. Eckert
Milton Helpern International
Center for Forensic Sciences
Wichita State University

INTRODUCTION

Forensic engineering is generally understood to mean the application of engineering principles and methodology to answer questions of fact that may have legal ramifications. The determination of the speed of a car at the time of a collision, the determination of the point of ignition of an explosion, or the assessment of adequacy of safety guarding are just a few examples.

By its nature, forensic engineering does not generally respect traditional disciplines. The analysis of an accident or failure often involves the application of several disciplines. For example, the investigation of a bridge collapse could involve the application of statics, dynamics, bridge architecture, corrosion chemistry, river hydraulics, soil mechanics and highway maintenance management.

As a general rule, an accident or failure is not the result of a single cause; it is usually the combination of several causes acting in concert or in sequence. An example of causes acting in sequence might be a gas explosion. The gas might have leaked from a corroded pipe, the corroded pipe may be the result of inadequate maintenance, and the inadequate maintenance may be the result of poor management practices. An example of causes acting in concert might be an automobile accident: both drivers may simultaneously take dangerous actions; one might be drunk;and the other's car might have a latent brake defect which becomes manifest only during hard braking.

Thus, the determination of the cause of such failures requires familiarity with a broad range of disciplines, and an ability to pursue several lines of investigation simultaneously.

The following chapters are intended to provide an introduction to the science, methodology and engineering principles involved in diagnosing some common types of accidents and failures. Each chapter stands alone and can be read by itself without reference to the others. The chapters have been written so that non-technical professionals involved in accident and failure analysis, such as insurance adjusters or lawyers, can gain a quick body of knowledge which can be immediately put to use. The chapters are also useful to technical professionals such as engineers and technicians, who may be unfamiliar with the particular investigative methodology or the technical points of interest.

In a modified form, some of the chapters were originally published as articles in *Forensic Ink*, a quarterly newsletter about forensic engineering, published by Dressler Consulting Engineers, Inc. Other chapters have been specifically written for this book. There is no rigid order of the arrangement of subjects; the selection of topics is based loosely upon the frequency of assignments and inquiries we have had at Dressler Consulting Engineers, Inc. In this regard, it is hoped that the topics will be of practical interest to the reader.

Where it is thought that the reader may wish to have further information, a small but very useful number of references are listed at the end of some chapters.

The chapters are intended as an introduction to the science and engineering principles used in diagnosing various types of accidents and failures. For this reason, mathematics has been kept to a minimum; however, a few algebraic expressions had to be included just to let the reader know that there is a hard-core analytical aspect to the subject.

Acknowledgments

In all endeavors such as this, there are many people who gave important support, without which the project would falter. Jean Solomonson and Gayle Robertson provided word processing. Judy Helms and Carol Fisher read proofs. Cindy Waller saw to it that things got done. Mike Hanson provided ideas for some chapters. Don Dressler, P.E., president of Dressler Consulting Engineers, Inc., offered suggestions and support for the work. I thank you all very much.

About the Author

Randall Noon is a licensed professional engineer, both in the U.S. and in Canada. Mr. Noon is employed by Dressler Consulting Engineers, Inc., in Overland Park, Kansas, and directs the company's failure and accident analysis activities. Mr. Noon holds a bachelor's degree in engineering from the University of Missouri-Columbia, a master's in engineering from the University of Missouri-Rolla, and a doctorate in engineering from Pacific Western University.

Warranties

No warranty is implied or expressed by the publication of this book. While every effort has been made to ensure accuracy and correctness, the publication is provided for instructional purposes only. Appropriate use of the materials presented is at the discretion of the reader, who assumes full responsibility.

Table of Contents

CHAPTER 1

Vehicular Accidents

INTRODUCTION

Automobile accidents happen frequently, and lead to numerous litigations with regard to medical expenses, disablement claims, property damages etc. **Accident reconstruction** is the usual term used to describe the application of scientific principles and methodology to determine the mechanics of an accident.

The mechanics of an accident include the velocities of the involved vehicles, the direction of their travel, any decelerations or accelerations that may have occurred, the sequence of events of the accident, forces upon the drivers, etc.

Accident reconstruction is primarily based on the actual physical evidence found at the scene or otherwise directly related to the accident. Scientific principles are then applied to analyze the accident. Generally, the application of these scientific principles will allow the various parameters to be directly calculated, or at least will establish upper or lower bounds to the possible values.

Statements or testimony of individuals are certainly taken into consideration, and important clues may often be contained in them that relate to the causation of the accident. However, a proper scientific analysis should be based upon verifiable facts and established natural laws, rather than on the biased opinion of an interested party.

1

Consider the following example:

Mr. Jones' car, which was travelling north in an intersection, was struck along its right side by the car of Mr. Wilson, who was travelling west in the same intersection. Mr. Wilson indicated that "his" light was green. Mr. Jones claims he entered the intersection just as "his" light turned amber, and that Mr. Wilson "jumped the gun".

Mr. Smith, a witness who was driving northward behind Mr. Jones, indicates that he saw Mr. Jones "peel out" when a light turned green and drive at least 70 mph to the next intersection to beat the light. Mr. Smith indicates that Mr. Jones ran a red light at the next intersection. When Mr. Smith reportedly saw Mr. Jones' car enter the intersection, Mr. Smith was even with a certain traffic sign, and Mr. Smith claims to have been going 35 mph at the time.

To "solve" this accident, the following factual information would be gathered at the scene:

1. Distance from sign noted by Mr. Smith to second intersection.

2. Distance from first intersection to the second intersection and lane widths.

3. The time duration from when the light turns green at the first intersection to when the light turns amber and then red.

4. The acceleration characteristics of Mr. Jones' car and of Mr. Wilson's car.

5. Tire marks in the second intersection indicating the side slip of Mr. Jones' car after impact with Mr. Wilson's car.

From the above information, the following could be calculated:

1. Whether Mr. Jones' car could reach a speed of 70 mph as suggested by Mr. Smith.

2. Whether Mr. Jones could reach the light before it turned amber or red.

3. At what speed Mr. Wilson could have struck Mr. Jones' car.

4. At what speed Mr. Jones' car was travelling when it was struck by Mr. Wilson's.

In short, the information in the statements of Messrs. Smith, Jones and Wilson can be compared to what is possible, what is reasonable and what actually fits the available information. Such a comparison will then allow the statements to be tested for veracity.

Accident reconstruction combines elements of the disciplines of kinematics, dynamics, material mechanics and energy methods, human factors engineering, automotive engineering and traffic engineering.

FALLS

One of the simpler accident situations to solve is a fall, such as might occur when a vehicle runs off a cliff or bridge. By measuring the height from where the vehicle fell, and the horizontal distance that it travelled before impact with the ground (or whatever), it is possible to ascertain the velocity of the vehicle just before it went off the cliff, with great accuracy.

For a simple fall, where the cliff edge is more or less horizontal, the following formulas hold:

$$t = \sqrt{(2hg)} \tag{1}$$

$$V_y = gt \tag{2}$$

$$V_x = x/t \tag{3}$$

$$V = \sqrt{V_x^2 + V_y^2} \tag{4}$$

where g = acceleration due to gravity
  h = altitude of fall
  x = horizontal distance covered during fall
  V_y= vertical speed when vehicle hit bottom
  V_x= horizontal speed, and
  V = resultant velocity at impact.

Figure 1. (1) Broken bolt eye on steering gear box caused steering failure. (2) Note damaged pinion. (3) Close up of pinion gear fracture.

The above formulas apply to a simple fall of a vehicle. The formula for "t" is used to determine the amount of time the vehicle took from fall to impact. The "h" value is the measured height of the fall. The "x" value is the measured horizontal distance from launch to impact. The value "V_y" is the velocity in the vertical direction upon impact, and the value "V_x" is the horizontal velocity of the vehicle when it left the top of the cliff. The value "V" denotes the vector sum of "V_y" and "V_x" which is the net velocity of the vehicle upon impact.

Sample Problem:

If a car ran over a cliff that was 34 feet high, and the car landed 52 feet along a horizontal line from the edge of the cliff, what was the speed of the vehicle when it left the top of the cliff?

Analysis:

By applying the formula for "t," it is found that it took 1.45 seconds to fall the 34 feet. (Use 32.17 ft/sec/sec for the value of "g.") When the vehicle hit the ground, it was falling at a vertical speed of 46.6 ft/sec, or 31.8 mph.

Because the vehicle impacted 52 feet from the edge of the cliff, its horizontal velocity when it left the top of the cliff was v(x) = 52 ft/1.45 sec = 35.9 ft/sec, or 24.5 mph.

The total, or resultant, velocity of the vehicle at the point of impact was 58.8 ft/sec, or 40.1 mph.

If the vehicle skidded or impacted upon some items prior to going over the edge, the speed of the vehicle may have been even greater. Additional analysis of those events might be required in some cases.

It is interesting that the above analysis was accomplished with essentially only two pieces of "hard" information - the height and horizontal travel.

SKIDS AND IMPACTS

When a vehicle is in motion, it has a certain amount of energy associated with its speed and mass. This type of energy is called **kinetic energy**. If the mass of the vehicle were magically doubled, the kinetic energy would be quadrupled. A vehicle at rest has zero kinetic energy.

The kinetic energy of a vehicle can be calculated by the following formula:

$$KE = (1/2)\ mv^2$$

where m = the mass of the vehicle
and v = the velocity of the vehicle.

In the English system of units, the kinetic energy is in units of pound force inches or pound force feet.

In order for a vehicle to come to rest, the kinetic energy of a vehicle must be dissipated. Under normal circumstances, that is the function of the brakes. The vehicle's brakes convert the kinetic energy of the vehicle into frictional work, which is accomplished by the rubbing of the brake pads against the wheel drums or brake disks.

In an accident, the kinetic energy of a vehicle can be dissipated by some of the following means:

1. Skidding and sliding (E_s).

2. Rollovers and flips (E_r).

3. Impact with other vehicles (crushing of the sheet metal around the vehicle (E_c).

4. Impact with stationary roadside objects.

5. Braking (E_o).

6. An increase in elevation (going uphill). This last item is a special type of energy term and is normally called **potential energy** (PE).

Thus, the following general equation can be used to describe an accident:

$$KE = E_s + E_r + E_c + E_o + PE$$

In a one-car accident, the "KE" term is for the one vehicle; in a two or multiple-car accident, the kinetic energy of each vehicle and its dissipation terms must be taken into account. Thus, several such equations may be needed, depending on the sequence of events.

The initial velocity of a vehicle, and most of the intermediate velocities at various points, can be determined by solving a specific energy equation of the type shown above. Essentially, the problem becomes one of accounting for all the ways the vehicles in the accident dissipated energy, and how much was dissipated in each process.

For example, the energy dissipated by skidding is given by the following formula:

$$E_s = Wdf$$

where W = the weight of the vehicle
 d = the skid distance, and
 f = the coefficient of friction between the road and the tires.

If a vehicle simply skids to a stop without hitting anything that would significantly slow it, the initial velocity can be calculated by the following formula:

$$V = \sqrt{2gdf}$$

Alternately, if only the initial velocity is known, the length of the skid can be determined by the following formula:

$$d = \frac{v^2}{2gf}$$

Sample Problem:

A car skidded and hit a pedestrian. The skid mark on the pavement was measured to be 62 feet long. How fast was the car going just prior to skidding?

Analysis:

Generally, striking a pedestrian does not significantly alter the speed of a typical car. Letting f = 0.7, and g = 32.17 ft/sec/sec, then substituting d = 62 feet into the formula indicates that v = 53 ft/sec, or 36 mph.

If the point of impact with the pedestrian were known with respect to the point of initiation of skid, the speed of the vehicle at impact could also be determined.

If there is uncertainty about the appropriate value of "f" to be used in the formula, a similar car could be deliberately skidded under similar conditions, and "f" could be determined experimentally.

The skid formula given above is often used inappropriately in accident analysis. The formula cannot be used correctly if any other type of energy dissipation event has taken place, such as a significant collision with another car. In such cases, the calculated speeds will be too low. If other types of energy dissipation have taken place, the kinetic energy (KE) general equation must be used instead.

CRUSH ENERGY DISSIPATION

In impact to the front of a vehicle, it has been found that the surrounding sheet-metal structure will be crushed in proportion to the amount of kinetic energy absorbed. In other words, the greater the speed impact, the greater the depth of crush.

Figure 2. Truck skid marks on highway.

Thus, by measuring the depth of crush impact damage, one can estimate the amount of energy absorbed by the vehicle.

In many types of accidents, an analysis of the length of skid and the amount of crush damage is enough to determine most of the accident parameters.

Because of this, it is important that good photographs be taken of the damaged vehicles. It often happens that accident reconstruction analysis takes place some time after the accident; sometimes years. In such cases, the engineer must rely on the police report and other "paper" records for information about the accident. A good set of documenting photographs of the accident scene and the vehicles will, upon proper analysis, provide the needed dimensional information.

In essence, various dimensions can be determined from photographic records by the same techniques used in drafting to construct two-point perspective drawings, albeit in a sort of reverse method.

While the amount of energy absorption needed to cause a certain crush depth may vary from vehicle to vehicle, it has been found that it is fairly consistent within a vehicle size class. For example, for compact cars, for each inch of crush to the front end, about 6,000 lb-ft of energy is absorbed. Other such factors can be applied to the rear end and sides of a vehicle, and to other classes of vehicles.

REFERENCES

1. *Motor Vehicle Accident Reconstruction and Cause Analysis* by Rudolph Limpert.
2. *The Traffic Accident Investigation Manual,* by Baker and Fricke.

CHAPTER 2

Vehicle Accident Reconstruction Using Basic Energy Analysis Techniques

INTRODUCTION

Information Available to the Engineer

The reconstruction of vehicle accidents can be a very difficult task. In most cases, the engineer will be asked to reconstruct the events of an accident long after the accident has occurred. Sometimes, the actual accident scene will be prohibitively far away from the engineer or will have changed by the time he is given the reconstruction assignment.

Relying upon the often conflicting information provided by witnesses or the accident participants can be confusing and misleading. Often, the witnesses will report their own conclusions and opinions instead of objective observations; sometimes the accident participants will knowingly or unknowingly lie about the events. Under these circumstances, obtaining factual information with which to work can be trying.

However, the engineer will usually have the following reasonably objective information available to him at the outset:

1. **The police accident report**. The police report will contain the usual basic identification information of the accident

11

participants. It will also note the position of the vehicles after the accident as found by the police, the location of skid marks, the point of impact, the general layout of the scene, weather and conditions data, and the general travel pattern of the vehicles before the accident.

2. **Photographs of the damaged vehicles.** This is usually available from the insurance companies involved or their adjuster agents.   They are used in evaluating insurance compensation to the accident participants.

The engineer may be asked to provide information or opinions about many aspects of the case, including some that are not related to the mechanical collision events. However, the engineer is nearly always asked to determine the initial velocities of the vehicles.

LIMITATIONS OF SOME EVALUATION TECHNIQUES

Accident events are often reconstructed using the basic classical laws of physics concerning force, velocity and motion. These physical laws provide a general framework for which events are possible and which impossible.   However, this strictly mechanical evaluation technique often has shortfalls in making exact determinations.

The application of strict kinematic techniques is often difficult. This technique requires knowledge of the velocities, accelerations, distances etc. of each vehicle from point to point in the accident sequence.

In classical physics, this technique is commonly applied to predict motion of a particle when all but one of the pertinent quantities are known. However, in a real accident there is generally more than one unknown variable. With several unknown variables, there are often not enough kinematic equations to provide a unique solution.

Thus, assumptions about initial or final velocities of the vehicles are often made in order to compute the other values. These assumptions are sometimes based on witness testimony or are just "good guess" estimates. This technique then becomes an iterative process where best guesses are made on some values until a good data fit is obtained.

The application of conservation of momentum techniques to the assessment of the accident sequence often adds considerable information. However it, too, requires knowledge of most of the initial and final velocities of the involved vehicles in order to fully satisfy or solve the momentum equations.

The application of dynamics, which involves the evaluation of forces along with the kinematic and momentum aspects, further advances the ability to assess an accident. However, dynamics is also a point-to-point sequence assessment technique. All or most of the forces must be determined at one point in order to determine the forces at the next point. Further, accident evaluation by dynamics will often not take into account the "lost forces" consumed in the deformation or destruction of the materials in the vehicles or at the scene.

ADVANTAGES OF THE ENERGY METHOD

The application of the energy method to the reconstruction of accidents has many advantages over the other techniques. In general, energy methods do not require detailed point-to-point interior analysis of the entire accident sequence.

Energy methods simply account for the energy required to create all the damage and effects observed which resulted from the accident. Knowing this allows calculation of the initial state of the accident; that is, the initial velocities of the vehicles.

In essence, the energy method is simply the application of the first law of thermodynamics: the energy in a closed system is conserved.

While dynamics, momentum, and kinematic evaluation techniques often require the use of free-body diagrams, vector algebra, and point-to-point evaluation, the energy method is scalar and simply requires knowledge of the accident end results.

Thus, the energy method is an excellent technique for accident reconstruction. All the basic information necessary to apply the energy method is normally contained in the police report and the adjuster's photographs.

THE FUNDAMENTAL PRINCIPLE OF THE ENERGY METHOD

The underlying principle of the energy method as applied to accident reconstruction is the first law of thermodynamics: the energy within a closed system is conserved.

However, in the reconstruction of vehicle accidents, the application of the first law may not be readily apparent.   Thus, the corollary is probably more useful: the work done on a particle by a force is equal to the change in kinetic energy of the particle.

In broad terms, this means that the initial kinetic energy of a vehicle (the energy it has by virtue of its velocity and mass) will be equal to the work that the vehicle does to come to a stop. This will include the work necessary to skid, change elevation, roll over, and crush metal.

As will be shown, the work components needed to skid, change elevation, roll over and crush sheet metal do not directly have velocity parameters in them. All four of the above common ways of converting kinetic energy into work can be measured by length: the length of the skid, the change in height of the center of mass during rollover, the depth of metal crush, and the change in the height of the center of gravity of the vehicle after an elevation change.   Thus, the work resulting from the accident is used to determine the initial velocities of the vehicles.

In general, the fundamental equation is as follows:

$$[KE] = [EC]+[Ef]+[ER]+[EE]+[EU] \tag{1}$$

where KE = kinetic energy
  EC = energy consumed by crushing
  ER = energy consumed by rollover
  Ef = energy consumed by skid friction
  EE = energy change due to elevation change
  EU = other types of energy consumption: sound, heat, etc.

KINETIC ENERGY

Work is defined as the distance through which a force will act upon a particle.

$$W = F(x) \tag{2}$$

where W = work
F = force
x = distance

It is also known that F = ma, where "m" is mass and "a" is acceleration.
If we wish to discuss an increment of work, then the following is true:

$$d(W) = F\ d(x) \tag{3}$$

where "d" signifies an infinitesimal increment.

Since F = ma = m(dv/dt) where "v" is the velocity of the particle, and "t" is time, then:

$$F\ d(x) = m\frac{dv}{dt}(dx) \tag{4}$$

Since v = dx/dt, then by rearranging:

$$F\ dx = mv(dv) \tag{5}$$

Finally then,

$$W = \int F(dx) = \int mv(dv) \tag{6}$$

By performing the integration, we obtain the equation for kinetic energy.

$$KE = W = (1/2)mv^2 \tag{7}$$

The change in kinetic energy of a particle between two different velocities is given by :

$$KE_1 - KE_2 = (1/2)m[v_1^2 - v_2^2] \qquad (8)$$

Note that the kinetic energy of a particle increases with the square of its velocity, but only linearly with its mass.

Example 1

Consider how much kinetic energy a 2,500-lb. car has when travelling at 20 mph:

$$KE = (0.5)(2500 \text{ lb}/[32.2 \text{ ft/sec}^2]) \text{ x } (29.3 \text{ ft/sec}^2)$$

Note: The weight of a vehicle is converted to mass by dividing by the gravitational constant, $g = 32.2$ ft./sec^2

$$mass = weight/(32.2 \text{ ft/sec}^2)$$

$$KE = 33,400 \text{ lb-ft.}$$

Example 2

Now consider how much reduction in velocity would occur if the same kinetic energy that the car has when it is going 20 mph is subtracted from the amount that it has when it is travelling 55 mph.

$$KE @ 55 \text{ mph} = 252,600 \text{ lb-ft}$$
$$KE @ 20 \text{ mph} = 33,400 \text{ lb-ft}$$

$$(1/2) mv^2 = 252,600 \text{ lb-ft} - 33,400 \text{ lb-ft}$$

$$v = 75 \text{ ft/sec, or } 51 \text{ mph}$$

Thus, if 33,400 lb.-ft. of kinetic energy were taken from a car going 20 mph, it would stop. But, if the same amount of kinetic energy were taken from the same car going 55 mph, the car would lose 4 mph.

FRICTIONAL ENERGY DISSIPATION

The resistance of an object to being moved along the ground is said to be proportional to the force or weight of the object against the ground. This proportionality constant, called the coefficient of friction, will vary from one type of surface to another. The work dissipated by pushing the vehicle across the ground is the weight of the vehicle times the coefficient of friction times the distance the vehicle was moved.

$$W = f \, s \, w \qquad (9)$$

where W = the work to overcome frictional resistance
      f = the coefficient of friction
      s = the distance of the skid
      w = the weight of the object

When a vehicle leaves rubber marks on the pavement or otherwise skids or slides over the ground, it is dissipating energy by friction between the ground and the tires.

If a vehicle comes to a stop by skidding alone, the general equation to determine the initial velocity is:

$$KE = Ef \qquad (10)$$

that is, the initial kinetic energy equals the energy dissipated by the skidding.

Substituting the given formulas for two energy terms:

$$(1/2)mv^2 = f \, s \, w$$
$$\text{and } v^2 = (2)(f)(s)(W/m)$$

Since w/m = g = 32.2 ft.sec^2, then

$$v = \sqrt{2 \, f \, s \, g} \qquad (11)$$

Note that the initial velocity of the vehicle is directly determined from the length of the skid mark on the pavement.

Some common friction coefficients are given in the following table:

Table 1. Some Typical Frictional Coefficients of
Automobile Tires on Various Surfaces

gravel and dirt road	0.35
wet grassy field	0.20
dry asphalt	0.60
wet asphalt	0.45
snow-covered road	0.20 - 0.30
ice	0.05 - 0.10
dry concrete	0.70
wet concrete	0.60

When a vehicle slides over several different surfaces, the energy dissipated is simply the sum of the frictional work done on the various surfaces.

Example 3

What was the initial velocity of a car that skidded first 50 feet on dry concrete and then 90 feet over a grassy pasture?

$$(1/2) \, mv^2 = w \, s_1 \, f_1 + w \, s_2 \, f_2$$

This algebraically simplifies to the following:

$$v = \sqrt{2g(f_1 s_1 + f_2 s_2)} \tag{12}$$

using $f_1 = 0.7$ and $f_2 = 0.2$. Then

$$v = \sqrt{2(32.2 ft/\sec -\sec) \times (0.7 \times 50 ft) + (0.2 \times 90 ft)}$$

and v = 58.4 ft/sec = 40 mph

Example 4

If a motorist applies his brakes after initially going 60 mph and skids 80 feet on dry concrete before crashing into a wall, at what speed would the vehicle have contacted the wall?

$$\frac{1}{2}m(v_1^2 - v_2^2) = f\,s\,w$$

Simplifying algebraically,

$$v_2 = \sqrt{v_1^2 - 2gsf} \qquad\qquad (13)$$

or v_2 = 64 ft/sec, or 44 mph.

Thus, the vehicle would have hit the wall with a velocity of 44 mph at the moment of impact.

It is notable that if the problem is only one of dissipating kinetic energy by frictional sliding on the ground or pavement, then the term for mass usually drops out of the final equation.

Example 5

A truck skids 100 feet on asphalt, turns over, and then slides another 200 feet further on its side (steel on asphalt, f = 0.15). How fast was it going prior to the initiation of skidding, and how fast was it going when it turned over?

Using Equation (12),

$$v_1 = \sqrt{2(32.2ft/sec/sec) \times (0.6 \times 100ft + 0.15 \times 200ft)}$$

v_1 = 76 ft/sec, or 52 mph initial velocity. Using Equation (13),

$$v_2 = \sqrt{v_1^2 - 2gsf}$$

v_2= 62 ft/sec, or 42 mph when the truck turned over.

In cases where there is rotation or yaw of the vehicle as it skids, the four skid marks of the tires (or however many there are) should be measured individually, and their mean length should be used for calculating the energy dissipation.

Further, the measurement of the skid length should be taken curvilinearly along the skid. If it is measured in a straight line from the first point to the last, without taking into account the extra length due to curvature, the determination of the initial velocity will err on the low side.

CHANGES IN ELEVATION

When a vehicle changes its elevation, it can either gain kinetic energy or lose it. Typically, if the vehicle goes uphill, kinetic energy will be lost as the vehicle works against the force of gravity. If the vehicle goes downhill, the downward acceleration from gravity will be converted into kinetic energy. More simply, a vehicle will skid further going downhill than going uphill.

In energy terms, this is a conversion of potential energy into kinetic energy and vice versa. Potential energy is energy that is stored and can be retrieved. In this case, potential energy increases with an upward change in elevation and decreases with a downward change in elevation.

The change in potential energy, the energy related to elevation, is given by the following:

$$[EE] = w(h_1 - h_2) \qquad (14)$$

where EE = energy associated with elevation
      w = weight of the vehicle
      h_1 = elevation at initiation of skid
      h_2 = elevation at end of skid, or where vehicle stops

When $h_2 > h_1$, then EE is a positive quantity, which means that the potential energy of the vehicle has increased, while the kinetic energy of the vehicle has been reduced by the same amount. Conversely, when $h_2 < h_1$, EE is negative, which means that the potential energy has been reduced and the kinetic energy has increased.

Considering only the energy terms associated with friction, kinetic energy, and change of elevation, then Equation (1) simplifies to the following:

$$[KE] = [Ef] + [EE] \qquad (15)$$

Substituting terms,

$$\frac{1}{2}m[v_1^2 - v_2^2] = w[h_2 - h_1] + [fsw] \qquad (16)$$

When $v_2 = 0$, and $h_1 = 0$, and the equation is solved for "v", the above equation simplifies to the following:

$$v = \sqrt{2gfs + 2gh} \qquad (17)$$

Example 6

A car is travelling 60 mph at point (1) and is coasting. Assume that there is no frictional dissipation. How high can it elevate itself above the initial elevation if it begins climbing a hill?

Using Equation (16) without the "skid" term, the equation then becomes:

$$\frac{1}{2}mv^2 = w[h_2 - h_1] \qquad (18)$$

Letting $h_1 = 0$ and solving for h_2,

$$h_2 = \frac{v^2}{2g} \qquad (19)$$

$$h_2 = 120 \text{ ft.}$$

Example 7

A car is travelling initially at a velocity of 40 mph. Its brakes fail totally, and it coasts 300 feet down a hill with a grade of 5 percent and crashes into a wall. How fast was the car going when it hit the wall? (Ignore friction).

Equation (16) is used to solve this problem. However, the skid term is ignored.

$$\frac{1}{2}m[v_1^2 - v_2^2] = w(h_2 - h_1) \tag{20}$$

$$v_2 = \sqrt{v_1^2 + 2g(h_2 - h_1)}$$

The elevation drop of a 5 percent grade over 300 feet is 15 feet. Then, h_2 = 15 ft, and h_1 = 0.

$$v_2 = \sqrt{3442 + 966} \ ft/sec$$

v_2 = 66 ft/sec, or 45 mph.

Example 8

A car skids 145 feet on dry concrete to a stop. At the end of the skid, the car is 12 feet lower in elevation than it was when the skid began. What was the initial velocity of the car?

Using Equation (17),

$$v = \sqrt{2(32.2 ft/sec/sec) \times (0.7 \times 145 ft - 12 ft)}$$

$$v = \sqrt{(64.4 ft/sec/sec) \times (101.5 ft - 12 ft)}$$

v = 76 ft.sec, or 52 mph.

In reviewing Example 8, it is worth noting how much over-all effect the elevation change factor made in the calculation. If there had been no change in elevation, the calculated velocity would have been 81 ft./sec., or 55 mph. If the 12-foot change in elevation was up (instead of down), the calculated velocity would have been 85 ft./sec., or 58 mph.

If the change in elevation in Example 8 had been an increase of 12 feet and the posted speed limit had been 55 mph, ignoring the

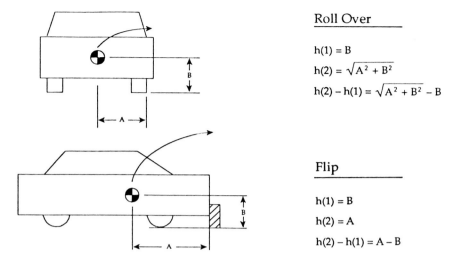

Roll Over _____

$h(1) = B$

$h(2) = \sqrt{A^2 + B^2}$

$h(2) - h(1) = \sqrt{A^2 + B^2} - B$

Flip _____

$h(1) = B$

$h(2) = A$

$h(2) - h(1) = A - B$

Figure 1. Vehicle rollover and flip geometry, showing center of gravity.

elevation factor would have allowed the conclusion that the car was travelling at the posted limit. Including the elevation factor would allow the conclusion that the car had exceeded the posted limit. Thus, the significance of the elevation term can be very important.

ENERGY DISSIPATION BY ROLLOVERS OR FLIPS

Sometimes a vehicle will roll over or flip over during a crash as a result of either hitting an obstacle or of centrifugal forces. When this happens, kinetic energy is converted to potential energy in a similar fashion as that which occurs with a change in elevation.

To calculate the energy dissipated by a rollover or flip, it is necessary to know or estimate the height of the center of gravity of the vehicle from the ground in the vehicle's usual position, and the maximum distance of the center of gravity from the ground during the flip or rollover.

In Figure 1, the rollover figure assumes that the car rotates about the outer edge of the tire and shows how to calculate the change in the height of its center of gravity.

In Figure 1, the flip figure assumes that the car rotates about the front edge of the car. A simple formula for the height change is also given for this situation.

In general, the formula for calculating the energy dissipated in a flip or rollover is as follows:

$$[ER] = w[h_2 - h_1]$$ (21)

where h_1 = the normal height of the center of gravity
    h_2 = the maximum height to which the center of gravity will be raised
    w = the weight of the vehicle
    [ER] = energy consumed by rollover

Equation (21) is similar in form to Equation (14), the change in elevation energy formula.

Typical values for changes in the height of the center of gravity for a medium-sized car are 1.5 to 2 feet for a rollover, and 4 to 5 feet for a flip.

During the rollover or flip, kinetic energy of the vehicle is converted to potential energy. As the flip or rollover is completed, the potential energy is theoretically converted right back to kinetic energy. In this respect, it might be thought that it is unnecessary to calculate the momentary loss of kinetic energy into potential energy since it will be restored in short order.

However, since the vehicle does not land back on its tires, the energy does not actually go back to kinetic energy, but is consumed in "wrinkling" the sheet metal during the roll. Thus it is necessary to calculate the energy lost by flip or rollover, because it is energy that is dissipated eventually in the sheet metal of the vehicle.

Example 9

A car weighing 3,000 lbs. skidded 75 feet on dry concrete and flipped over after hitting a low retaining wall. The car then skidded on its roof another 235 feet on dry asphalt. How fast was it going initially?

Using f = 0.7 for the concrete skid, and f = 0.10 for the skid by the roof on the asphalt, and assuming that the change in center of gravity height during the flip was 6 feet, the initial velocity is calculated as:

$$v = \sqrt{2g(s_1 f_1 + s_2 f_2 + h)}$$

$$v = \sqrt{64.4(0.7 \times 75 + 0.1 \times 235 + 5)} \ ft/sec$$

$$v = \sqrt{64.4(52.5 + 23.5 + 5)} \ ft/sec$$

$$v = 72 \ ft/sec, \ or \ 49 \ mph.$$

If the energy dissipated by the flip were ignored, the calculated velocity would be 70 ft./sec., or 47.7 mph. Consideration of the flipover then made a difference of just over 1 mph in the calculation of the initial velocity.

ENERGY DISSIPATION BY CRUSHING

When two bodies collide, two kinds of deformation occur. The first type of deformation is described as elastic; this is often a temporary deformation similar to the action of a spring. A spring can be compressed by an applied force and thus store energy. However, when the spring is released, it will regain its original shape and reconvert the stored energy to kinetic energy.

The second type of deformation is termed **plastic** or **residual**. In plastic deformation, the material does not spring back. This is the kind of deformation that is most conspicuous on a vehicle after an accident; crumpled and irregularly folded sheet metal.

The graph in Figure 2 shows how typical steel will elongate when subjected to a tensile force. Note that the steel initially goes through an elastic region and then goes through a plastic region. If the steel is not stretched into the plastic zone, it will snap back to its original position. If, however, it is stretched into the plastic zone, it will not snap back all the way and will have permanent or residual deformation.

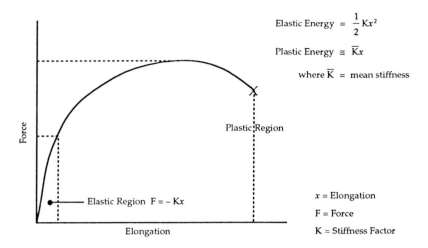

Figure 2. Plastic and elastic energy in deformed steel.

In the elastic zone, the relationship between force and deformation will be linear and is given by the following:

$$F = k\,u \qquad (22)$$

where k = the spring or stiffness constant
u = the amount of deformation

Equation (22) is sometimes referred to as Hooke's Law.

The work done to cause elastic deformation is derived as follows:

$$dW = F(du) = (ku)(du) \qquad (23)$$

where du = an incremental distance
dW = an incremental amount of work

$$W = \int F(du) = \int (ku)du = \frac{1}{2}ku^2 \qquad (24)$$

In equation (24) the work term W, which represents the work needed to cause the material to elastically deform, is similar to a potential energy term. After the work is applied to cause the deformation, it is possible to retrieve the work.

During the contact portion of a collision, some of the initial kinetic energy of the vehicle is converted into elastic energy in the form of elastic deformation of the vehicle. However, the elastically deformed material will "spring back" after the collision is complete. When the material springs back, its elastic energy is retrieved and usually is converted back into kinetic energy. This is why vehicles will often rebound slightly from the point of impact.

This is what occurs when a steel ball bearing is dropped onto a solid concrete floor. The ball hits the floor, is momentarily squashed, and then regains its shape and springs back up into the air.

On the other hand, when one drops a ball of mud onto the floor, it does not bounce back; it flattens out on the floor. This is an example of a purely plastic deformation. The kinetic energy of its initial velocity is converted into permanent material deformation.

The work done in causing plastic deformation is lost; it is not retrievable like elastic energy. Thus, once energy has been used to cause plastic deformation, it becomes unavailable for conversion back into kinetic energy.

The work associated with plastic deformation is approximated within certain limits by the following:

$$W = (Fc) \, u \qquad\qquad (25)$$

where "Fc" is a force constant, or stiffness factor, similar to a frictional force. In vehicles, it is called the **crush constant**.

In experiments involving a 1980 Chevy Citation, front-end crush was measured at three impact speeds against an immovable barrier. The impact speeds were 35 mph, 40 mph, and 48 mph. The crush distances were 21.4 inches, 27.9 inches, and 40.4 inches. The weight of the car was 3130 lbs. Calculating the respective kinetic energies and dividing by the crush distance to obtain Fc factors, the following was calculated:

for 35 mph   Fc = 5985 lbf.-ft./in.
for 40 mph   Fc = 5996 lbf.-ft./in.
for 48 mph   Fc = 5962 lbf.-ft./in.

The above shows that within the range of speeds tested, the stiffness factor or crush constant has a variation of less than one percent.

Thus, it is possible to determine the amount of energy absorbed by a vehicle when it is damaged by simply measuring the depth of crush and multiplying it by the appropriate crush constant.

Due to basic design, most vehicles will have a different stiffness or crush constant for the front, sides, and rear of the vehicle. In general, a car will be stiffer on the sides of the vehicle and least stiff in the rear of the vehicle.

Table 2 lists stiffness factors calculated from actual front-end crash data of a number of cars.

Table 2. Front end stiffness, or crush constants

VEHICLE	Weight (lb.)	Crush Constant (lb.-ft./in.)
Small Cars		
1979 Honda Civic	2180	4720
1979 Ford Fiesta	2190	4040
1979 Plymouth Champ	2310	4260
1979 Datsun 210	2430	3960
1979 VW Rabbit	2600	4860
1979 Toyota Corolla	2650	5340
1979 Chevette	2730	5150
Average	2441	4619
Range		+16/-14%
MEDIUM CARS		
1979 Mustang	3070	7610
1979 Mercury Capri	3070	7178
1979 Chev. Monza	3240	5970
1977 Volvo 242	3290	4600

VEHICLE	Weight (lb.)	Crush Constant (lb.-ft./in.)
1979 Ford Fairmont	3300	6000
1982 Volvo DL	3350	5040
1979 Volvo 244DL	3370	4960
Average	3241	5908
Range		+28/-22%
FULL SIZED CARS		
1980 AMC Concord	3700	7460
1979 Plymouth Volare	3820	7170
1979 Olds Cutlass	3820	5600
1979 BMW 528	3840	6400
1979 Ford Granada	3950	6145
1979 Mercury Marquis	4220	6300
1979 Ford LTD	4370	6850
1979 Dodge St.Regis	4460	6470
1979 Olds 98 Regency	4710	7355
1979 Ford LTD II	4810	6000
1979 Lincoln Cont.	5360	7384
Average	4278	6649
Range		+12/-16%

Inspection of Table 2 shows that the stiffness factor increases generally with the weight of the vehicle. In general, the accuracy of using an average stiffness factor for a weight category of vehicles (excepting Volvos) for a vehicle with an unknown stiffness factor is +/- 16 percent. In this respect, a +/- 16 percent accuracy in computing the crush energy allows a +7/-8 percent accuracy in computing the vehicle's initial velocity.

Table 3. Miscellaneous Average Crush Constants

Side Crush Factors for average car	9,000 - 11,000 lbf.-ft/in
Rear Crush Factors for average car	3,500 - 6,000 lbf.-ft/in
Side Crush Factors for average car hit by motorcycle	200 lb m.c 1,200 lbf.-ft/in
	350 lb m.c 1,750 lbf.-ft/in
Wheel Base Reduction factor in motorcycles	1,150 lbf.-ft/in

Example 10

A 1979 Citation skids 90 feet and impacts with a wall head-on. It also rebounds from the wall leaving an additional 5-foot skidmark. The crush distance is measured to be about 16 inches. How fast was the car initially travelling before the initiation of the skid?

$$\frac{1}{2}mv^2 = w[fs_1 + fs_2] + (Fc)u$$

$$v = \sqrt{2g[fs_1 + fs_2] + (\frac{2}{m})(Fc)u}$$

Assuming f = 0.7, Fc = 6000 lbf-ft/in, and w = 3130 lbs, then

v = 79 ft/sec, or 54 mph

When determining the initial velocities of two separate vehicles that collide, an energy analysis of the accident will be able to determine the combined total initial kinetic energies of the two vehicles. In order to calculate the individual initial velocities, it is necessary to determine a ratio between the two initial velocities. Such a ratio can be determined by considering the momentum balance of the accident and the associated coefficient of restitution by establishing one of the speeds from kinematic considerations, or by determining that the relative kinetic energy of one of the two vehicles is negligible.

When two bodies collide, they will first deform. At some moment during this contact they will have the same velocity. When this happens, the two bodies will have the same force applied between them for the same duration of time. The formula for force times time, $F(dt) = m(dv)$, indicates then that the two bodies will have the same impulse:

$$m_1[v(1,1)-v] = m_2[v-v(2,1)] = I(def)$$

where I(def) = impulse during deformation.

After the maximum deformation has occurred, the two bodies will then spring back to some degree. This is termed **restitution**. During restitution, the two bodies will again share the same applied force between them over the same time period. Thus,

$$m_1[v-v(1,2)] = m_2[v(2,2)-v] = I(res)$$

where I(res) = impulse during restitution.

The ratio of I(def) to I(res) is called the coefficient of restitution and is a measure of the elasticity (or plasticity) of the collision.

The coefficient of restitution is then defined as follows:

$$e = \frac{[v(2,2) - v(1,2)]}{[v(1,1) - v(2,1)]}$$

where the first number in the velocity parentheses indicates the vehicle, and the second number indicates either before the collision at time (1) or after the collision at time (2).

If a collision is purely elastic, the coefficient of restitution will be 1.0; if it is purely plastic, it will be 0. For moderate car velocities of about 25 mph, "e" will be approximately 0.2. For higher velocities, "e" will be equal to 0.05 to 0.1.

Example 11

Car "A", which weighs 3,500 pounds, travels south. Car "B", which weighs 2,400 pounds, faces north. The cars collide head-on. After the collision, car "A" does not move from the point of impact, and car "B" skids backwards (south) for 8 feet.

Car "A" has a front-end stiffness of 5,000 lbf-ft/in and is crushed to a depth of 25 inches. The coefficient of friction is 0.55. Determine the initial velocities.

By equation (10), the velocity of car "B" after the collision is calculated as follows:

$$v = \sqrt{2gfs}$$

v = 17 ft/sec.

The energy equation for the accident will be as follows:

$$\frac{1}{2}m(A)v(A)^2 + \frac{1}{2}m(B)v(B)^2 = [w(B)]fs + Fc(A)u(A) + Fc(B)u(B)$$

Substituting the known quantities,

$$v(A)^2 + 0.686v(B)^2 = 283 + 1840 + 1840 = 3963 \; ft^2/sec^2$$

By trial and error, a number of velocity combinations could be listed to satisfy the above equation. For example: if v(A) were 44 ft/sec, then v(B) would be 54 ft/sec. If v(A) were 60 ft/sec, then v(B) would be 23 ft/sec.

However, if it is assumed that the collision was relatively elastic and has a coefficient of restitution of 0.27, then:

$$e = 0.27 = v/[v(A) - v(B)]$$

Since v = 17 ft/sec, then v(A)-v(B)=63 ft/sec.

Combining the above relationship of v(A) to v(B) with the energy equation allows a solution to be found.

$$v(A) = 63 \text{ ft/sec and } v(B) = 0 \text{ ft/sec.}$$

The above solution indicates that while car "A" was traveling south at 43 mph, car "B" was stopped.

Example 12

A large truck weighing 75,000 lbs. travels north. A car crosses in front of the truck traveling east to west, and is hit in the side by the truck. The truck skids 21 feet before impact with the car. The car and truck skid together on concrete pavement for 97 feet. After this, the truck turns on its side and slides another 154 feet. The car then slides through dirt in the median strip for 89 feet. The car rolls over once. The car is crushed in the side about 24 inches. The truck bumper and lower fender on the right side of the front is crushed about one foot. What was the speed of the truck?

To solve the problem, the following is assumed:

$f = 0.7$ for tires of truck and car on concrete
$f = 0.1$ for side of truck on pavement
$f = 0.3$ for tires of car on dirt
$w(1) = 75,000$ lbs. for truck
$w(2) = 2,500$ lbs. for car
$h = 1.6$ ft, rollover change in height, car
$h = 2.0$ ft, turnover change in height, truck
$Fc(1) = 17,000$ lb-ft/in for truck front
$Fc(2) = 9,100$ lb-ft/in for car side

From kinematic considerations, it is known that the car could have a velocity of no more than 25 mph, or 37 ft/sec.

Since there are many terms to be considered in this problem, it is useful to make a table for the ways the energy was dissipated:

Energy Dissipation Terms

truck skid
  0.7 (75,000 lbs)(21 ft) = 1,103,000 lb-ft

truck and car skid
  0.7(77,500 lbs)(97 ft) = 5,262,000 lb-ft

truck slide on side
0.1(75,000 lbs)(154 ft) = 1,155,000 lb-ft

truck turnover
(75,000 lbs)(2 ft) = 150,000 lb-ft

car slide in dirt
0.3(2,500 lbs)(89 ft) = 67,000 lb-ft

car rollover
(2,500 lbs)(1.6 ft) = 4,000 lb-ft

car crush in side
(24 in)(9,100 lb-ft/in) = 218,400 lb-ft

truck crush in front
(12 in)(17,500 lb-ft/in) = 210,000 lb-ft

unaccounted-for energy = E

total dissipated energy > 8,170,000 lb-ft

The kinetic energy of a car weighing 2,500 lbs. traveling at 25 mph is 52,200 lb-ft. This is less than one percent of the total energy dissipated and will therefore be neglected.

Thus, the speed of the truck was as follows:

$$\frac{1}{2}mv^2 = Total\ E$$

FUTURE ACCIDENT RECONSTRUCTION PROBLEMS

In the near future, vehicles will be equipped with computer-controlled braking systems and suspension systems. The goal of such systems is better braking traction and vehicle handling.

It has been known for some time that the maximum braking effect does not occur when the wheels are locked and the vehicle skids, but occurs when the wheels are allowed to roll just enough to avoid skidding. The coordination of the suspension system with the braking system via a microcomputer sensor and control system can, theoretically, greatly improve the handling of a vehicle in emergency maneuvers and avoid loss of vehicle control.

However, these computer-controlled braking systems will likely leave different and less distinguishable skid marks prior to a collision. If the system remains intact after a collision, post-collision skid marks may be hard to find also.

Thus, while such a control system may be a boon to the driver, it could deprive the accident reconstruction engineer of useful information about the positions of the vehicles during the course of the accident and when the application of hard braking took place.

This means that with no skid marks with which to work, the engineer will have to rely more heavily on the crush damage and final position of the involved vehicles to determine direction, velocity, and the other pertinent accident parameters.

It is possible to apply energy dissipation factors for the brakes themselves, assuming a factor for kinetic energy dissipation per length of vehicle travel. However, it will be more difficult for the engineer to independently verify when such maximum braking began and ended. This will make the mathematical determination of initial velocities prior to the initiation of braking difficult and possibly wholly indeterminate without using witness testimony.

Further, the introduction of a variety of composite materials into the automobile frame structure will further complicate the calculation of plastic deformation energies. In the past, mild steel was the universal material for vehicle structures. Coupled with the similarity of automotive frame design, reasonably accurate prediction of crush depth could be made of untested vehicles.

However, a wide variety of new materials are likely to be gradually introduced in the coming years. Such materials may include graphite composites, plastics, fiber reinforced metals and plastics, new aluminum alloys, ceramics, metaloceramics and reinforced foils. Many of these materials will have very different characteristics than the traditional mild steel and will likely be very different from each other as well. Also, it is likely that the car companies will introduce these new materials in different parts of the frame at different times.

In short, the explosion of automotive innovation in materials, controls, and design in the next decade will require the accident reconstruction engineer to be familiar with a larger data bank of experimental information and be able to apply a greater number of theoretical techniques to achieve solutions.

REFERENCES

1. Limpert, *Motor Vehicle Accident Reconstruction and Cause Analysis,* Second Edition, Chapters 16, 17 and 28, Michie Company, 1984.
2. Hight, Hight and Lent-Koop, "Barrier Equivalent Velocity, Delta V, and CRASH3 Stiffness in Automobile Collisions", *Field Accidents: Data Collection, Analysis, Methodologies, and Crash Injury Reconstructions,* SAE P-159, 1985.
3. Baker and Fricke, *Traffic Accident Investigation Manual,* Northwest University Traffic Institute, 1986.
4. Stonex and Nelson, *Collision Damage Severity Scale*, SAE, 1970.
5. Grimes and Jones, *Assessing Equivalent Barrier Speeds in Accident Investigations,* 1972.
6. Bhushan, *Analysis of Automobile Collisions*, SAE 750895, October 1975.

CHAPTER 3

Application of the Classical Scientific Method to Accident Reconstruction

DEFINITION OF TERMS

Accident reconstruction is a term used in the field of forensic engineering. It refers to the determination by study and analysis of the sequence of events which resulted in an accident or failure.

For example, in a two-car accident, a forensic engineer might examine the damage on the two cars, visit the accident scene and note any marks or damages, review testimony, review police reports etc. From this body of information, he would then determine the sequence of events in the accident, and calculate any related quantitative information such as impact speeds, pre-braking speeds, directions of motion, etc. When this is completed, the accident is said to have been "reconstructed".

Forensic engineering itself is the application of engineering science and methodology to answer questions of fact related to legal proceedings. Typically, forensic engineers are called upon to reconstruct accident events and failures, and to reveal their findings to a court of law or an arbitration board for consideration in its deliberations.

ORIGINS OF THE SCIENTIFIC METHOD

The formal origins of the scientific method are often credited to Aristotle, in his elucidation of the inductive method of reasoning (1). In this method, a general rule is established, based upon the evidence of many individually observed facts or instances that demonstrate an underlying commonality. In this way, basic principles or universal propositions which are not self-evident or readily apparent may be discovered.

A possible pitfall of inductive reasoning, as also discussed by Aristotle (2), is that the number of observations may be too small for a true generalization to be made. A false generalization can be made when only the observations of a limited set are considered.

While Aristotle succinctly established the basis for the modern scientific method, he did not generally practice it. In general, the Greek school of thinking did not stress experimentation to verify scientific "truths" (3). Greek scientific thought was mostly centered around *a priori* reasoning, where the underlying truths or principles are assumed from logical "first principles", in the same way that geometric proofs are constructed.

Often, complicated logical constructs had to be invented to reconcile actual observations with the assumed first principles. For example, it was assumed as a first principle that circles were a perfect geometric form, and that the earth was the center of the universe. Because heavenly things such as the stars and the planets were also perfect, then, it was reasoned that the stars and the planets must revolve around the earth in perfect circles. It was not necessary to experimentally verify this, because it was obvious to any one who could see that this was correct (notwithstanding Aristarchus of Samos, of course).

This deduction, formalized by Ptolemy, led to a remarkably complex system of circles, circles within circles, and circles within those circles, to explain the observed motions of the planets (4).

Of course, this method of scientific inquiry led to many other incorrect perceptions of the world. However, due to the general appeal of the method and to its similarity to Christian doctrine, the method became dogma in Western thought. The *a priori* methodology was not seriously challenged until the advent of the Renaissance some 1500 years later.

DEVELOPMENT OF THE SCIENTIFIC METHOD

Promulgation of the modern scientific method is generally credited to Roger Bacon, a thirteenth-century Franciscan monk (5). Roger Bacon was educated at Oxford and at the University of Paris, and became interested in scientific studies, especially experimentally based inquiries. These pursuits caused his superiors to doubt his religious conviction, and he was accused of black magic. For ten years he was confined to a Paris monastery and was denied all writing materials, instruments and books.

Upon the accession to Papacy of Pope Clement IV, Roger Bacon was again allowed to work. However, his scientific writings were banned. When Pope Clement IV died, Bacon was again confined for another ten years. In his later years he returned to Oxford.

His major work was *Opus Majus*, in which he argued for a reformation of the sciences, and specifically for the practice of experimentally based science. *Opus Majus* was eventually published in 1733, centuries after his death in 1294.

The basis for objection to the experimentally based type of scientific inquiry, as recommended by Roger Bacon, was that the senses (i.e. observations of experiments) were not to be trusted since the senses could be tricked in many ways, including by deviltry.

It was also argued that fundamental religious tenets must always come first. Therefore, some things must naturally be assumed *a priori*. The formal reasoning methodology used in the Church was to be the basis for all inquiry into the nature of things, whether of heaven or of earth. To do otherwise was viewed as a threat to the logical underpinnings of the Church.

It was for these reasons that the heliocentric theory of the solar system, published by Copernicus in 1543, was a milestone with respect to the modern scientific method: this was the first time that a major scientific theory was proposed solely because it was consistent with observational data.

Figure 1. Sequence of photographs showing collapse of grain elevator tube in Oakley, Kansas in 1983: (1) Note dust cloud at top. (2) Tube blows out at bottom where structural stresses are greatest. (3) Collapse of upper tube and outflow of grain. (4) Collapse complete. (5) View of the damage. (6) View from the air; note train stopped in right of photo, avoiding spilled grain pile.

THE SCIENTIFIC METHOD

The modern scientific method, as discussed by Nobel laureate Dr. Linus Pauling (7), is briefly as follows: from a number of observations a general statement is derived which correlates with all observed facts; this is a **hypothesis**.

The hypothesis is then tested by making extensive additional observations or experiments. Importantly, the experiments are often designed to prove that the original hypothesis is wrong. When new facts are discovered that are not explained by the hypothesis, the hypothesis must be modified or discarded, and a new hypothesis is proposed.

A hypothesis becomes a **theory** when it fully accounts for all the observed facts. Over time, as the theory is successfully used predictively, and has been found free of errors, it is given the status of a scientific **law**.

THE SCIENTIFIC METHOD APPLIED TO ACCIDENT RECONSTRUCTION

With respect to accident reconstruction, the scientific method is slightly modified to account for the fact that a single event is being analyzed, an event which is unlikely to be repeated.

The scientific method, as applied to accident reconstruction, is as follows: A general working hypothesis is proposed, based upon "first cut", verified information. As more information is gathered, the original working hypothesis is modified or changed to encompass the growing body of observations.

After a certain time, the working hypothesis can be tested by using it to predict the presence of evidence that may not have been obvious or was overlooked during the initial information-gathering effort.

A hypothesis is considered a complete "accident reconstruction" when the following are satisfied:

1. The hypothesis accounts for all the verified observations.

2. The hypothesis accurately predicts additional evidence not apparent initially.

3. The hypothesis is consistent with accepted scientific laws and methodology.

SHORTCOMINGS

As noted by A.J. Ayer in discussing the *a priori* method versus empiricism, "No matter how often it (a general proposition) is verified in practice, there still remains the possibility that it will be confuted on some future occasion" (8).   This has particular relevance to accident reconstruction, because it is unlikely that an accident will occur again in exactly the same way for additional study to be made.

In the modern scientific method, it is usual to design experiments where the principle being studied is not obscured or complicated by other principles acting simultaneously. The variable being studied is usually singled out to be free from other influences.   Then the experiment is repeated a number of times under varying situations to ensure that the hypothesis holds for all conditions.   Numerous outcomes consistent with the hypothesis provide a statistical basis for the validity of the hypothesis. In fact, an estimate of the possible error of the hypothesis can generally be made, based upon the number of experimental tests (9).

In accident reconstruction, while similar accident events can be compared to the one under study, they are usually not exactly the same, and may not be free from other influences; they are not single-variable studies.   Further, it is often the case that an accident event cannot be repeated many times for study and statistical confirmation. This can be due to cost, safety considerations, or because the exact conditions existing at the time of the accident are not completely known.

However, it can be argued that if there is a great body of observed facts about an accident event, this is a suitable substitute for experimental statistical confirmation.   Only the "correct" accident reconstruction hypothesis can account for all of the observations.

An analogous example might be the determination of an algebraic equation from a plot of points on a graph. The more data points there are on the graph, the better the curve fit will be. Theoretically, a large body of data points with excellent correspondence to a certain curve would give proof that the fitted curve was equal to the original data generator function.

The reality of accident reconstruction, however, is that the accident event itself may destroy evidence about itself. Important evidence can be lost or obscured in the accident debris. In short, there can be observational gaps.

Using the graph analogy, this is similar to having areas of the graph with no data points, or with few data points. With fewer points available to define the curve, there is a wider variety of curve combinations that could be fitted to the points available.

Thus, it is possible that the available observations can be explained by several accident reconstruction hypotheses. Gaps or paucity in the observational data may not allow a unique solution. In effect, two qualified and otherwise forthright experts could proffer two conflicting accident reconstruction hypotheses, both equally consistent with the available data.

THE LEGAL SYSTEM

Having several plausible explanations for an accident event may not necessarily be adverse. In the legal system, it can sometimes happen that one does not have to know exactly what happened, but rather what did **not** happen.

For example, suppose that the speed limit is 35 mph, and it is desired to know if a motorist was exceeding the speed limit. The available observation data may not allow an exact determination of the vehicle's speed, but it may provide enough information to prove that it was greater than 50 mph. Thus, while a unique solution is not possible, a solution sufficient for legal requirements may be feasible.

In our adversarial legal system, one person or party is the accuser, or plaintiff, and the other is the accused, or the defendant. The plaintiff is required to prove that the defendant has done some wrong to him. However, the defendant has to prove merely that he did not do the wrong; he does not have to prove who or what did.

Thus, even if the observational data is not sufficient to provide a unique accident reconstruction solution, it may be sufficient to deny a particular one.

A PRIORI BIASES

One of the thornier problems in accident reconstruction is the insidious application of *a priori* methodology. This occurs when legal counsel hires a forensic engineer to find out only information beneficial to his client's position. The counsellor will not specifically state what findings are to be made, but may suggest that since the "other side" will be giving information detrimental to his client, there is no pressing need to repeat that work.

While this argument may at first seem innocuous enough, it serves to bias the original data because only beneficial observations will be considered, and detrimental ones will be ignored. If enough "bad" evidence is ignored,, the remaining observations will eventually force a beneficial accident reconstruction. Like a graph analogy, if enough data points are erased or ignored, almost any curve can be fitted to the remaining points.

A second variation of this a *priori* problem occurs when a client doesn't provide all the observational data base to the forensic engineer for evaluation. Important facts are held back. This similarly reduces the observational data base, and enlarges the number of plausible hypotheses which might explain the facts.

A third variant of insidious *a priori* methodology is when the forensic engineer becomes an advocate for his client. In such cases, the forensic engineer assumes that his client's legal posture is true, even before he has evaluated the data base. This occurs because of friendship, sympathy, or a desire to please his client in hope of future assignments.

To guard against this, most states require that forensic engineers accept payment only on a time and materials basis. Unlike legal counsellors, forensic engineers may not work on a contingency basis. This, at least, removes the temptation of reward or a share of the winnings.

Further, it is common for both the adversarial parties to hire an accident reconstruction expert. During court examination by the attorneys for each party, the judge or jury can decide for themselves whether the expert is biased. During such court examinations, the terms of hire of the expert are questioned, his qualifications are examined, any unusual relationships with the client are discussed, all observations and facts he considered in reaching his conclusions are questioned, etc.

While an expert is considered a special type of witness due to his training and experience, he is not held exempt from adversarial challenges. While not perfect, this system does provide a way to check such biases and *a priori* assumptions.

SUMMARY

The scientific method is the foundation of modern science. This method is based on observation, repeatable experimentation, and the application of inductive logic. A modified version of the scientific method is the foundation of accident reconstruction.

However, in accident reconstruction, the application of the scientific method is often limited to a single, unrepeatable catastrophic event, and the points of concern are not limited to a single variable. Further, the accident itself may hide or destroy part of the observational data base. These limitations may not allow a single solution; several alternate accident reconstructions may be consistent with the available data. Additionally, it is possible for various *a priori* biases of the forensic engineer to influence the conclusions of an accident reconstruction.

To provide a check against these potential deficiencies, the adversarial nature of the legal system allows competing expert theories to be presented, and the experts themselves to be examined for relevant competency and possible biases. The court can then decide for itself which assessment os the best fit for the circumstances.

REFERENCES

1. Aristotle, *On man in the Universe, Introduction* by Louside Loomis, Walter Black, Inc., Roslyn, N.Y., 1943, page xiv.
2. *Ibid*, page xv.
3. *The Columbia History of the World*, edited by Garraty and Gay, Harper & Row, New York, 1981, pages 681 - 686.

4. *Ibid*, page 685.
5. *Funk and Wagnalls Standard Reference Encyclopedia*, Vol. 3, Reader's Digest Books, 1970, page 897.
6. *The Columbia History of the World*, edited by Garraty and Gay, Harper & Row, New York, 1981, pages 685 - 686.
7. *General Chemistry* by Linus Pauling, Dover Publications, New York, 1970, pages 13 - 15.
8. *Reason and Responsibility* edited by Joel Feinburg, "The A Priori", by A.J. Ayer, Dickenson Publishing, 1971, page 181.
9. *Introduction to Mathematical Statistics* by Paul Hoel, Chapter 8: General Principles for Statistical Inference, John Wiley & Sons, New York, 1971, pages 190 - 224.

CHAPTER 4

Motorcycle Accidents

INTRODUCTION

Motorcycle accidents occur at about 2.6 times the rate of automobile accidents, on a per-mile-driven basis. This chapter examines some of the reasons for this; analysis is given to explain the difficulty of judging the distance and speed of an approaching motorcycle at night because of its single headlight, and also the apparent lack of visual perception by other drivers during daylight.

There were about 181 million registered cars, trucks and buses in the U.S. in 1987. Of this total, 139 million were automobiles. In addition, there were also some 5.1 million registered motorcycles. In terms of total registered vehicles, motorcycles accounted for about 2.7 percent, cars 75 percent, and trucks 22.6 percent of the total.

The average motorcycle travelled 1,789 miles a year in 1986, and the total motorcycle road miles driven were 9,400 million. The average car, bus or truck travelled about 10,157 miles in the year, and the corresponding total road miles were 1,840,000 million. In general, the average registered motorcycle was driven only 17.6 percent as far as the automobile, truck, or bus, and the total road miles for motorcycles were only 0.5 percent of that of cars, trucks and buses.

Considering these statistics, one might expect that the number of accidents involving motorcycles would be correspondingly low. All

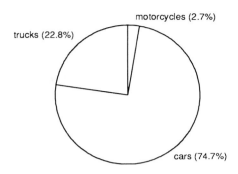

Figure 1.

other factors being equal, the number of accidents would be a function of road miles driven, since the primary risk of accident exists while driving. Based on the miles driven, then, we would expect that 0.5 percent of all accidents involved motorcycles.

The total number of car accidents in 1986 was 27.7 million; the total number of truck accidents was 6.1 million, and motorcycle accidents was 440,000. Of the total of 34.2 million vehicle accidents, car accidents accounted for 81 percent, truck accidents 17.8 percent, and motorcycle accidents 1.3 percent.

The number of deaths resulting within 30 days of a vehicular accident in 1986 was 46,100. Persons in vehicles accounted for 33,700, pedestrians for 6,800, motorcyclists 4,600, and bicyclists 900. Of the deaths related to all vehicular accidents, motorcyclists accounted for 10 percent of the total.

In short, the statistics show that motorcycles have 2.6 times more accidents than other vehicles, and 20 times more fatalities per road mile driven.

The reason for the dramatically higher fatality rate for motorcycles is readily apparent: in almost all severe collisions, the motorcyclist is ejected from his motorcycle and must dissipate his or her kinetic energy by impacting against the ground or some other object, often another vehicle. In this respect, the fatality rate per road mile driven simply reflects the relative safety in an accident of motorcycles against other types of vehicles.

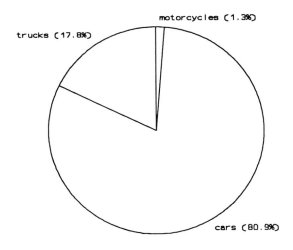

Figure 2.

However, the reason why motorcycles are involved in more accidents per road mile driven is not as readily apparent. Some may argue that a motorcycle should be better able to avoid an accident, due to its greater maneuverability. Some may argue that since fewer infirm and feeble persons drive motorcycles than cars, the average motorcyclist should have better reflexes and would therefore react faster in accident situations.

In my personal experience, a large proportion of the motorcycle accidents I have analyzed involve motorcycles driving straight ahead which have struck a car making a left-hand turn of some kind. This may be a left turn into a driveway or into a cross street. In nearly all of these cases, the driver of the car stated that he or she never saw the motorcycle, or did not see it until it was too late. Thus, the perception of the motorcycle or motorcyclist by drivers of other vehicles appears to be a key factor in the relatively high accident rate of motorcycles.

Some of the perception factors that contribute to the higher per-mile-driven accident rate of motorcycles are explored below.

PERCEPTION OF HEADLIGHTS

The next time you drive at night, notice how you perceive the speed and distance of an oncoming vehicle. If it is a car, truck or bus, you first observe the angular separation of its two headlights, and this will indicate how far away the vehicle is. Since the distance between the left and right headlights is more or less standardized, an experienced driver can estimate the distance from the oncoming vehicle by the amount of angular separation between its headlights. To estimate the closing speed, observe for a few moments the rate at which the angular separation between the headlights increases.

For example, my daughter's 1975 Ford Mustang has headlights that are 4' 4" (132 cm) apart. At 500 feet (152 m), the angular separation between the headlights is 0.497 degrees. At 250 feet it is 0.993 degrees, and at 125 feet, 1.985 degrees. The following figure plots the angular separation of the 1975 Mustang's headlights against its closing distance.
Mathematically, the angular separation is given by:

$$a = \arctan x/d \tag{1}$$

where x = distance from center of left headlight to center of right headlight,
d = closing distance between cars, and
a = angular separation of headlights in degrees.

Due to the small angles under consideration and the closing distances involved, the angular separation is for all practical purposes a linear function as follows;

$$a = 57.3 \ (x/d) \tag{2}$$

Even as close as 50 feet, the error of the linear approximation given in equation (2) is less than 1 percent.

Thus, when the headlights are twice as far apart as first observed, the oncoming car has halved the closing distance. This linear relationship is easy and appeals to our common intuitive notions of speed and distance. For most purposes, the small variation in headlight spacing among car models is negligible.

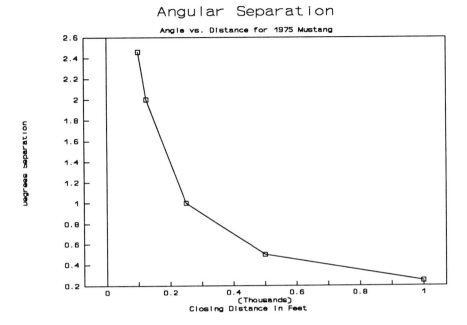

Figure 3.

However, a motorcycle has only one headlight; thus, the angular separation of headlight will not cue a driver of the distance and closing speed of an approaching motorcycle. The only cue available is the intensity of its headlight.

Light falls off in intensity as the square of the distance. In other words, a light with an intensity of 24 footcandles at 50 feet will have an intensity of 6 footcandles at 100 feet. Mathematically the relationship is as follows:

$$I = \frac{K}{d^2} \tag{3}$$

where K = an arbitrary constant,
      d = closing distance, and
      I = intensity of the light.

In order for a driver to estimate the closing distance of an approaching motorcycle, he or she must be able to compare the light intensity of the headlight as he or she sees it with some memorized value. This is very difficult for many reasons. Some types and brands of headlights are brighter than others. The aim of the headlight will affect the brightness perceived by the driver, and the slope and angle of the road may also affect how the headlight is aimed at the driver. Further, road dirt on the headlight may affect its brightness, as well as the presence of fog, snow, rain, smoke and other atmospheric conditions.

Further, in order to correctly estimate the speed of an approaching motorcycle, the driver must compare the intensity if its headlight at one instant with another. In this situation, the increase in intensity will vary as the square of the closing distance.

Compare the equation for a two-headlight car with that of a one-headlight motorcycle:

for car,

$$a = \frac{k}{d} \tag{4}$$

for motorcycle,

$$I = \frac{K}{d^2} \tag{5}$$

where k,K = arbitrary constants,
   a = angular separation of headlights,
   I = headlight intensity, and
   d = closing distance.

Thus, in judging the distance and speed of an oncoming motorcycle, a driver must use a different and less reliable cue and apply a different mathematical rule to estimate speed.

As noted in the introduction, only 2.7 percent of registered vehicles were motorcycles, and motorcycles accounted for only 0.5 percent of all miles driven. Thus, most drivers are very experienced in observing two-headlight cars, trucks and buses at night, but motorcycles are far less frequently seen. In cold climates, motorcycles may not even be encountered at all during winter months.

Thus, as compared to cars, trucks and buses with two headlights, a driver is more likely to misjudge the closing distance and speed of a motorcycle because:

1. The visual cue is different.

2. The rule that applies to the cue is different.

3. The experience factor in applying (1) and (2) is low.

4. The driver may mistake the approaching single headlight for a car or truck with a burned-out headlight. The driver may then mentally "fill in" the area around the single headlight in an attempt to assign a position to the phantom car or truck.

DAYLIGHT PERCEPTION

The maximum width of a 1975 Mustang, and of most cars, is about 72 inches. The maximum width of a motorcycle without extensive fairing is about 30 inches. Thus, the width of the car or truck is generally about 2.4 times that of a motorcycle.

A person with 20/20 vision will be able to readily distinguish and read 4-point type at a distance of 12 inches. A 4-point type letter is 4/72 inches tall. Thus, a person with 20/20 acuity can distinguish and read type that subtends an arc of 1/4 degree, which is about half the width of a full moon.

A person with 20/50 vision can readily distinguish and read materials at 20 feet, that a person with 20/20 vision can readily distinguish and read at 50 feet. Many states allow persons with vision correctable to 20/50 to drive a vehicle.

For our purposes, let us assume that a person with 20/20 vision will fully recognize a vehicle during daylight hours when it subtends a horizontal arc of at least 1/4 degree. For a car, this will occur when the car is 1,375 feet away - about a quarter mile. For a motorcycle, this will occur when the motorcycle is only 573 feet away.

A person with 20/50 vision would not be able to distinguish the car until it subtended a horizontal arc of 5/8 degree; this occurs when the car is 550 away, or when a motorcycle is 229 feet away.

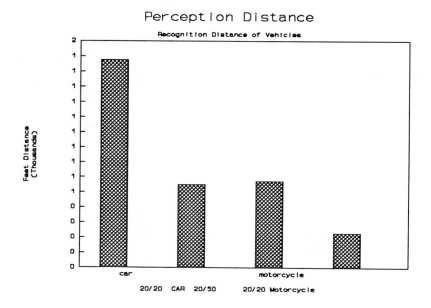

Figure 4.

Most two-lane highways are posted for no more than 55 mph, which is about 80.7 ft./sec. If two vehicles approach each other at the maximum posted speed, their relative closing velocity is 110 mph or 161.4 ft./sec. Thus, a driver with 20/50 vision would discern an oncoming car 3.4 seconds before they meet, but would discern a motorcycle only 1.4 seconds before meeting it.

It is well known that younger people have faster reflexes than older people. A reaction time of one-half second for young drivers, and one second or more for older drivers is not unusual. If both an oncoming motorcycle and car were going 55 mph, an older driver with 20/50 vision may barely have enough time to recognize it and react before the vehicles meet.

As previously mentioned, my own experience has found a significant portion of car/motorcycle accidents involve an oncoming motorcycle and a left-turning car. Since a modest car is 16 feet long, a left turn from a stop will take 19 feet of arc for the front end of the car to reach the left shoulder, and another 16 feet to get the car clear of the roadway. A moderate acceleration for a medium car from stop is 5 ft./sec./sec. Thus, the time needed for a modest car to fully execute a left turn and be clear of the left lane is about 3.74 seconds.

If we assume that the driver has 20/50 vision, he or she will discern a car at 550 feet, and a motorcycle at 229 feet away. This means that an oncoming car can be approaching at 147 ft./sec. (100 mph) or less, and will be seen in time for the driver to avoid making a left turn in front of it. However, an oncoming motorcycle could collide with the left-turning car if the motorcycle were exceeding 61 ft./sec. (42 mph). The driver may not see the motorcycle in time before making the left turn, and the motorcycle may close the distance before the car can fully execute the turn.

The problem is further exacerbated by the fact that during hard braking, motorcycles require more technique from the rider, especially while turning. In fact, even under normal conditions, a motorcycle typically takes longer to stop than a car. Thus, while the motorcyclist may see the car well in advance, he may not perceive that it is about to make a left turn until he is too close to execute a complete stop.

At a typical "hard" braking deceleration of 16 ft./sec./sec., a motorcycle needs 2.76 seconds to come to a complete stop from an initial speed of 42 mph. This does not allow for any reaction time. If a motorcyclist travelling at 42 mph does not realize within one second that the car is initiating a left turn in front of him, the motorcyclist will not be able to fully stop before hitting the car. The motorcyclist must then make an evasive turn of some sort to avoid impact with the car.

Thus, while the above kinematic analysis shows that drivers with 20/20 vision will likely be able to see a motorcycle well in advance of a problem, a driver with 20/50 vision may not discern it in time. It may be no coincidence that the width of a car is about 2.5 times that of a motorcycle, and that the rate of motorcycle accidents is 2.6 times that for other motor vehicles.

REFERENCE

1. Traffic Accident Data, *Statistical Abstract of the United States*, 1989.

CHAPTER 5

Photographic Documentation of Accident-Damaged Vehicles

ACCIDENT DOCUMENTATION

Often an engineer is asked to analyze a vehicular accident long after the accident has occurred. In such cases, the vehicles may have been sold long since as scrap, and the original accident scene may have been changed by a highway upgrade project. If this happened, the accident analysis must be done from available documentation. Primarily this will consist of the police accident report and the adjuster's file information.

Unfortunately, police accident reports vary greatly in quality from place to place due to differences in training and experience. Further, important details may be overlooked due to the pressures of caring for injured parties, dealing with traffic flow, and removing debris from the scene. Some important details are often overlooked because at the time of the accident, no one would have anticipated their importance.

Because of this, any photographic documentation accumulated by the adjuster is likely to become an important resource for reconstructing an accident. As a matter of routine, it is common for an adjuster to take some photographs of the vehicles involved, if only to document their damages for insurance coverage issues.

In some cases, the adjuster may be the only person who had time to look at the actual vehicles and accident scene in any detail. If there is litigation several years later, a few well-chosen photographs can make a crucial difference in the outcome of the case.

CHOICE OF CAMERA

It is recommended that photographs be taken with a 35-mm camera, using 200 or 400 ASA speed color film.   Good 35-mm photographs can be enlarged, made into slides for effective presentation, and examined under magnification for details.   The negative allows for many copies to be made, and sometimes development tricks can be used to highlight or bring out details that may not have otherwise been visible in the prints.

Generally, the use of black-and-white film is not recommended. Color can be very useful in differentiating an oil stain from a water spot, a rust spot from dirt, a grease streak from a mud streak, etc. If necessary, black-and-white prints can be made from color negatives.

The use of instant "Polaroid"(©) type film and cameras is not recommended; their photographs tend to have poor detail resolution. They are more difficult to enlarge or make into slides, and when they are enlarged, the results tend to be fuzzy. Sometimes their colors fade after several years. Since development occurs on the spot and there are no permanent or useful negatives, there is no opportunity to lighten or darken prints to bring out detail, and reprints tend to be a problem.

Unfortunately, many companies and police departments provide their employees with instant-type cameras and film because of the convenience of being able to immediately staple a photograph of the damaged vehicle to the file.   However, the proliferation of one-hour photo development shops has made even this advantage questionable.

PHOTOGRAPHS

In using photographs to document information, it is important to remember two rules:

> Film is much cheaper than travel; when in doubt, photograph it.

> Take photographs as if you were a draftsperson preparing drawings of the accident particulars.

With respect to the second point, when a good set of photographs is available, it may be unnecessary to even take written notes. All the information will be in the photographs. While the accuracy of a person's written notes may be called into question, a clear photograph of the same detail is difficult to dispute.

ITEMS TO PHOTOGRAPH

Photograph all four sides of the vehicle; and if possible, do a top view as well. Photographs should be like the views of a shop drawing. Similarly, like special features in a shop drawing that bear noting in a larger-scale detail, special accident details should be photographed close up.

In this way, it is an easy matter to factor dimensions of unmeasured items in the photograph by comparing them to known items. For example, it might be important to know how deep the front end was crushed in a collision. Analysis of the top, right side and left side views can provide that dimension even if the actual crash measurement was not taken.

Photograph the tires to show their tread depth and condition. Be sure they are the actual tires the car was using; sometimes replacement wheels and tires will be put on a vehicle to facilitate towing. Photograph the insides of the wheels; brake cylinder failures normally result in a "wet" spot on the inside of the wheel drums. Photograph the level of the brake fluid in the master cylinder. Many types of brake failures cause loss of fluid.

Photograph the dashboard and controls to show their positions. It may be important to know that the radio and air conditioner were

Figure 1. Photo of wheel shows car ran on rim before hitting shoulder.

on and that the windows were up. Under such circumstances, the driver might not be able to hear a train whistle or the horn of another vehicle. Likewise, it may be important to know whether the lights or the wipers were on, etc.

In stuck-accelerator cases, it is useful to know the position of the gear selector at the time of the accident. Also, photograph the gas pedal, brake pedal and driver's floor mat. Their relative size and position can be important.

Photograph the driver and passenger spaces. The presence of spilled drinks or food, open maps, cassette tapes, etc., in the car may indicate that the driver was distracted. The presence of liquor containers or drug vials has obvious importance.

Photograph the driver's door edge. Generally, the driver's door will have the manufacturer's tags and ID and oil-change and maintenance stickers; these items can be very important in establishing maintenance dates. The manufacturer's ID can be used to trace any applicable recalls.

Photograph the seat belts. Often in an accident, crush damage will pin the seat belts in their last position. This may be useful in establishing whether they were being worn at the time of the accident.

When there is a question about whether or not headlights or turn signals were on, take close-up photographs of the light filaments in the lamps, or whatever is left of them. When a lamp is on, a severe impact will cause its filament to distend. When a lamp is on and its lens is broken, the filament will burn and leave a yellowish residue on the reflector.

Photograph the fuse block. This may be useful in establishing whether or not certain electrical items were operable at the time of accident.

ACCIDENT SCENE

Photograph each vehicle's approach to the point of impact from the driver's point of view, being sure to include any warning signs, speed limits, etc. Interestingly, a common mistake made on police accident reports is an incorrectly listed speed limit. It is useful to do this at about the same time of day the accident took place, soon after the accident, so that the shadows and sunlight are about the same. For example, it might be important to know if the sun was directly behind the only traffic light at the time of the accident.

Photograph accident pavement marks if they still exist. Radiator spots, oil spots, etc., are useful in determining exactly where the vehicles may have been located at various times. Due to the difficulties in determining scale on pavement close-ups, it is useful to put something in the photograph like a ruler or even a standard blank sheet of paper.

When one of the vehicles has struck stationary objects, it will often leave a "tag" mark, where a streak of paint from the vehicle rubbed off onto the object. Finding these tag marks is useful not only in establishing the trajectory of the vehicle, but in establishing which vehicle hit what object. This is especially true with concrete barriers and guard rails.

ANALYSIS OF PHOTOGRAPHS

The technique for analyzing accident photographs is basically "reverse" projective geometry. In projective geometry, a draftsperson is given a set of dimensions for a part, and is then asked to draw the part as it would appear from a particular location or perspective. With photographs, the view is given first and the dimensions are to be determined.

A very simple example would be a photograph of a brick wall. If a person wanted to know how high the wall was, he would only have to count the number of brick courses in the photograph. Since a brick course has a standard thickness, the height of the wall would be the number of courses times the usual thickness of a brick course.

Suppose a car drove into a brick wall, and its front end was crushed to an unknown depth. Suppose also that the only evidence of the crush is a photograph of the right side of the vehicle after it was damaged. If the make and model of the car is identifiable, then the original over-all car length and the tire diameter dimensions are available in published references. Within the photograph, the crushed length of the car can be measured in tire diameters. The length of the car after the accident would then be its length in tire diameters times the tire diameter. The crush depth would then be the over-all length of an undamaged vehicle less the measured length of the damaged one.

In this respect, an adjuster does not have to painstakingly make a lot of measurements to document the vehicle or accident scene. He only has to take the appropriate photographs; all the needed information is then in the photographs.

CHAPTER 6

Vehicular And Marine Lighting

ON/OFF LIGHT ISSUES

Numerous car and boat accidents occur at nighttime. In such cases, it is not unusual for one party in the accident to claim that the headlights, taillights, running lights, etc. of the other party were not on at the time of the accident. Of course, if the lights in fact were not on, there is a legitimate issue concerning the visibility of the unlit vehicle or boat, and the prudence of the person operating it.

Even when the accident has not occurred at night, there can be reasons for wanting to know if certain lights were on. For example, in some cars the backup lights will come on whenever the car is put in reverse. If it can be shown that the backup lights were on at the time of the accident, the position of the gear selector can be determined.

Similarly, a common cause of motorcycle-car accidents is the failure of motorists to see the smaller motorcycle. Newer motorcycles are wired in such a way that their headlight will turn on automatically whenever the motorcycle is running. This is done to improve the motorcycle's visibility (and meet many state laws). However, if the headlight were to burn out a fuse, it might be possible to operate the motorcycle with the headlight not working. If it can be shown that the headlight of a motorcycle was off at the time of an accident, then negligence on the part of the motorcyclist might be demonstrated.

FILAMENT MATERIAL

Commercial incandescent light filaments are nearly universally made of tungsten. This is a relatively heavy metal, with an atomic number of 74. It falls into the same periodic family as chromium and molybdenum. At room temperature, tungsten is brittle and hard, and has a gray appearance.

One of tungsten's attributes, with respect to light filaments, is that it has the highest melting point of all the metals - 6,170 degrees Fahrenheit. It is electrically conductive and can be drawn into a fine wire. This combination of characteristics makes it an excellent material for incandescent light filaments. In an incandescent light bulb, electric current passes through the filament and causes it to heat up enough to emit visible light. Unlike most metals, tungsten has melting point higher than the temperatures at which incandescence occurs, which varies between 4,000 and 5,500 degrees Fahrenheit.

OXIDATION

Tungsten will oxidize in air at only 752 degrees Fahrenheit, to form tungsten oxide (WO_3). Since this is well below the temperature at which incandescence occurs, in order to keep the filament from simply burning up, it is encased in a bulb filled with a nonreacting gas such as nitrogen. While tungsten's reactivity with oxygen may seem like a complication with respect to the manufacture of light bulbs, it can be useful in determining whether a light was on at the time of an accident.

If the bulb around a filament is broken open, so that air can reach the hot filament, the filament will literally burn with a yellow flame, and the resulting product, tungsten oxide, is a yellow powder that does not dissolve in water. It is, however, soluble in strong alkalies.

Thus, we may conclude that a light was on at the time of an accident if:

1. A yellow flame was observed at the light just after the accident.

Figure 1. Distention in both filaments.

Figure 2. Distention of remaining filament.

2. A yellow powder or a whitish smoke stain is observed on the reflector, in the bulb remains, or elsewhere around the light.
3. The filament wire is noticeably darkened.

Chemical analysis of the powder or smoke residue can further confirm that the material is tungsten oxide. In this regard, the identification of tungsten oxide by chemical or physical means is a relatively unique marker. There are few common uses for free tungsten other than light filaments, and the only relatively common use of tungsten oxide is in yellow pigment in ceramics. Unless a person were to crash a car into a store specializing in yellow ceramic pots, the identification of tungsten oxide near a broken light is a very certain indication that the light was on when the bulb envelope was broken.

Figure 3. Broken but unstretched headlight filament; note pieces of coiled filament scattered in glass envelope.

Further, the fact that tungsten oxide is insoluble in water means that rain will not wash it away; tungsten-oxide smoke stains on a headlight reflector can be observed months after the accident.

In some accident cases, it might be claimed that the lights were on, but that the accident severed the wiring before the bulb broke open. Thus, it is reasoned, the filament would not burn. However, this is a specious argument.

The oxidation temperature of tungsten is well below the temperature range of incandescence. When a light bulb is turned off, it takes some time for the filament to cool down below its oxidation temperature. Since most accidents are measured in seconds, or fractions of seconds, there is generally not enough time for a lamp filament to cool down enough to avoid any oxidation effects.

Also, if the bulb is relatively cool with respect to the oxidation temperature, or little air reaches the filament before it cools, the filament coil or wire may simply exhibit discoloration in proportion to its temperature. Dark colors in the wire, such as purple, indicate higher temperatures than lighter colors, such as yellow.

Of course, if the bulb envelope is broken open, the lack of any tungsten oxide formation or darkening indicates that the filament temperature was below the oxidation point, and that the light had been off at least a short while prior to the accident.

BRITTLENESS

At ambient temperatures, tungsten is very brittle and exhibits no practical ductility. In effect, the material yield point and the rupture point are the same; i.e., the material does not stretch before it breaks. This characteristic can be put to good use to determine whether the lights were off, or at least "cold" at the time of an accident.

In accidents where the impact deceleration is high, the filaments may break apart. If they break apart with no discernible stretching of the filament coils, the filament was "cold", i.e., the light was not on.

Since a "cold" filament does not burn in air, this characteristic can be applied without regard to whether the bulb envelope is broken or intact, and is useful in corroborating a lack of tungsten oxide formation.

DUCTILITY

At elevated temperatures, tungsten becomes ductile. The following table shows the relation of temperature to true plastic strain in tungsten.

Table 1. True Plastic Strain vs. Temperature
for Commercially Pure Tungsten

Temperature	True Stress	True Strain
600°F	65 ksi	0.1 in/in
800°F	65 ksi	0.2 in/in
1000°F	65 ksi	0.3 in/in
1200°F	65 ksi	0.4 in/in

The temperature at which tungsten changes from brittle to ductile is about 645 degrees Fahrenheit. This is not very different from the temperature at which is oxidizes in air.

If an impact occur when the light is on, and the bulb envelope is not broken open, it is possible to still confirm that the light was on, due to ductile stretch of the filament coil. The filament will simply stretch out in the direction of deceleration.

In fact, by testing several light bulbs of the same manufacturer at various acceleration levels, it is possible to roughly calibrate the stretch of the filament, so that the magnitude of the peak deceleration can be determined by the relative stretch of the filament. In cases where other sources of information are scant, this could be useful in analyzing the accident. The direction of stretch is also a useful secondary marker to corroborate the impact vector. However, it should be noted that "rebound" effects in the filament, and multiple impacts which are tightly spaced together, may introduce some uncertainty in interpreting this effect.

A similar calibration can be done for cold filaments, but it yields only a single deceleration threshold. Either the deceleration exceeds the filament break level, or it does not.

In cases where there are two filaments in one bulb, such as in a headlight or certain tail light, the lit filament will indirectly heat up the unlit filament due to proximity. Thus, the lit bulb will stretch the most, and the unlit bulb will stretch the least or perhaps break. Such markers can be very useful in establishing whether the high beams were on, instead of the low beams.

TURN SIGNALS

Turn signals normally come off and on one or two times a second. While in a signal mode, therefore, the cooling-down time of a turn signal is no more than about 0.5 seconds. Oxidation of a signal light filament can still occur up to a second or so after the filament is turned off, and darkening and stretching can still occur several seconds after being turned off. All of the markers used in steady-light type bulbs can be applied to turn signals and flashers, to determine whether they were on at the time of an accident.

OTHER APPLICATIONS

The filament stretching effect can be used to determine if any of the interior lights of a vehicle were on at the time of the accident, such as the dome light, the map light, license plate light, etc. In boating accidents, the same principle can be applied to determine which of the running lights were on at the time of a collision.

MELTED GLASS

When a glass bulb envelope breaks open, there are usually many tiny shards produced. If some of these shards land on the hot filament, they may partially melt and stick to the filament. Examination under low magnification is usually required to observe this effect. Finding such melted glass particles on a filament can confirm that the light was incandescent when the glass was broken.

SOURCES OF ERROR

In cases where a vehicle has experienced a previous impact, the lights may exhibit stretch from the previous accident.

Occasionally, after an accident has occurred, a person may turn on lights in an uninformed attempt to see what still works. This can cause an otherwise unburned filament in a cracked bulb envelope to burn up or darken.

In very low-velocity impact accidents where the bulb is not broken, there may not be enough impact deceleration to cause filament deformation.   In such cases, the lack of markers does not signify anything.

REFERENCES

1. *The Traffic-Accident Investigation Manual*, Baker and Fricke, Northwestern University Traffic Institute, 1986, Section 23.
2. *Atlas of Stress-Strain Curves*, ASM International, 1987, pages 577-579.

CHAPTER 7

Storm Lightning Damage

BASIC LIGHTNING FACTS

In the areas of Missouri, eastern Kansas, eastern Oklahoma and northern Arkansas, there are, on average, between 50 and 60 days a year when thunderstorms exist in any given locality. However, this is relatively light in comparison to certain parts of Florida, which may have over 100 thunderstorms a year.

In the Kansas City area, for example, there were 59 thunderstorms in 1986. Translated, this means that in 1986 there was a 16-percent chance of a thunderstorm on any given day. Thus, if any piece of equipment were to fail due to random wear and tear, it would have a 16-percent chance of failing on a day during which a thunderstorm took place.

In some months, the probability of a thunderstorm occurring on any given day of the month is significantly higher. In Kansas City in June 1987, there were 13 days with thunderstorms. Thus, on any day in June 1987 there was a 43-percent chance that a thunderstorm occurred.

Given that all equipment, including electrical equipment, will eventually fail due to wear and tear, some equipment is bound to fail

71

on a day in which there is a thunderstorm. This is, of course, a circuitous way to make the point that simply because a piece of equipment fails on a day in which a thunderstorm occurs does not automatically imply that it was damaged by lightning.

This points up two significant aspects of lightning damage:

1. Lightning damage cannot occur unless there is a thunder storm, and
2. The mere occurrence of a thunderstorm in the area does not prove that lightning damage took place.

Thunder is heard no more than about 16 miles from the originating lightning strike. Further, most lightning occurs from cloud to cloud, rather than from cloud to ground. It is estimated that 85 percent of all lightning is cloud to cloud. Thus, even reports of thunder in an area are not proof that lightning was close enough to have caused damage.

In short, the only way to verify lightning damage is to examine the equipment and determine if the damage noted is consistent with lightning characteristics.

LIGHTNING DAMAGE

Lightning is essentially a discharge of static electricity which has built up enough charge potential to overcome the breakdown resistance of the intervening air gap.

The potential differences in a typical strike about 1.2 miles long range from 100 to 1,000 million volts. The average peak current equals approximately 20,000 amperes. Peak electrical power in a typical strike occurs within 1 to 10 microseconds.

Usually more than one discharge occurs in a strike. Lightning which occurs nearby is usually accompanied by a thunderclap that will "knock a person's socks off".

Lightning can cause three types of damage to electrical equipment: thermal, electrical, and electromechanical. Typical damage caused by lightning might include fractured electrical components, burned components, disheveled windings, vaporized printed-circuit conductors and blown terminals.

Figure 1. Lightning damage to printed circuit board. Arcing was from board to board.

VERIFYING LIGHTNING

The first step in verifying possible lightning damage is to accurately determine the date and time on which the damage took place. This will allow for an accurate comparison against weather records.

Generally, a repairman will have diagnosed the damage in question as having been caused by lightning. The repairman should be asked to specify which tests or observations led him to the conclusion that the damage was caused by lightning, rather than by normal wear and tear. A simple observation that a unit doesn't work any more is not credible evidence; neither is the often-repeated assertion, "I've seen lightning before, and this is it."

Were a car diagnosed as having a bad fuel pump, we would expect that the mechanic would have the fuel pump to show us for our inspection. Similarly, if a repairman has diagnosed lightning damage, he should be able to produce lightning-damaged parts for proof and examination.

THE TWO-BOARD SHELL GAME

Unscrupulous repairmen have been known to do the following: after they have been called in to fix a computer or telephone system, they indicate that several computer boards are severely lightning-damaged, and need immediate replacement with new ones. They then substitute used boards and charge new-board prices, expecting that few people will check that the replacement boards are in fact not new.

Figure 2. Lightning damage inside a power-supply cabinet. Note pattern of arcing to metal door.

The repairman then reports that the old boards are beyond repair and says that he threw them away. However, he may have actually sent them to the factory for repair. He will either get a salvage value refunded to him, or will simply resell the repaired boards for new ones again.

WEAR, TEAR OR LIGHTNING?

The "chips" that make small computers and computerized telephone systems possible are actually very delicate components. Most chips operate at 5 volts and are easily damaged by higher voltages.

Electronic devices using chips have a power supply that converts the regular house service of 115 volts AC into the necessary DC voltages to power the circuit boards. Most power supplies do not contain circuitry to "buffer" the output voltages; thus, changes in the input voltage to the power supply cause corresponding changes in the output DC voltages supplied to the boards.

The power provided by utilities is not considered "computer clean"; the line voltage will contain momentary spikes and blinkouts. These spikes or blinkouts will cause damage to the chips over a period of time, and will hasten their eventual failure.

The difference between the utility-line surges and lightning-type surges is in their magnitude and duration. A 10-microsecond pulse caused by lightning will do much greater damage than the smaller nanosecond pulses caused by switching relays in the utility lines. Typical utility surges will be perhaps two, three or four times the normal voltage; lightning surges will typically be measured in thousands of volts.

During a thunderstorm, the utility-generated surges increase in number due to the increased switching activity. Power-utility surges can occur because of trees falling on power lines, car accidents with power poles, damaged utility equipment, fried squirrels, and other non-lightning causes. During these increased periods of surge activity, electronic chips suffer an increased risk of failure caused by the "pounding" they take from these voltage transients.

If out-of-the-ordinary surges have occurred in the utility lines, the utility company will often have a record of it. Records will also show whether any utility equipment was struck by lightning. If the records do not indicate that a surge has occurred in the monitored lines, the remaining possibilities for origination of a surge are limited.

REDUCING LIGHTNING RISKS

Most of the surges which damage electronic equipment come through the power service of a building. A measure of protection against such surges can be achieved by using surge protectors or arrestors.

The use of lightning arrestors in utility equipment has been standard practice for years, and is one reason why, with the exception of direct hits, utility equipment does not have the same rate of lightning failures as do small computers and telephone systems.

A good surge protector will detect surges in the line and will shunt them to ground before they can damage the equipment. Some surge protectors will protect an item from typical line spikes and blinkouts inherent in utility-provided power, but will not protect it against lightning-type surges. Some will protect against both types of surges.

However, most products sold as surge protectors are not very good and offer little practical protection. The buyer should be careful in choosing one: it is wise to ask either the local utility or the company from which the electronic equipment was purchased as to an appropriate protection device.

The operation of modern electronic equipment containing chips without using surge protection is unwise and greatly increases the chance of an early failure. Unfortunately, it is not unheard of for a $2-million CAT-scan machine to be significantly damaged by a lightning surge which could have been averted by a $500 surge protector.

REFERENCES

1. Martin Uman, *Lightning*, Dover Publications, New York, 1969.
2. Ivars Peterson, *In Search of Electrical Surges*" Science News, Vol. 132, December 12, 1987, pp. 378-379.
3. Daniel Waxler, *Modeling the Effects of Lightning on Electrical Equipment*, U.S. Army Armament Research and Development Command, ARCL-TR-78029, November 1978.

CHAPTER 8

Vibration and Blasting Damage

AIR BLAST VIBRATIONS

Vibrations from nearby blasting are typically transmitted by air or by ground. Damage caused by air vibrations, usually referred to as the blast concussion or the airborne shock wave, occurs when the blast is relatively close to the structure.

If the effect of topography or intervening structures are not present when a blast occurs, the blast creates a concussion or airborne shock wave which radiates in all directions. The effect is similar to the waves which radiate out of the point where a stone drops into a pool of water. Like the water waves in our analogy, the shock wave will arrive and contact the side of the structure that faces the point of origin of the blast.

The weakest structural element of a building is, typically, the windows. Loosely set glass will often break out when subjected to a pressure difference of only 1 lb./sq.in. Typical windows will often break out when the pressure difference is 2 lbs./sq.in. As a comparison, it takes a wind speed of over 200 mph to produce a pressure of 2 lbs./sq.in.

At the pressure thresholds necessary to break glass, the associated sound-wave level exceeds 90 decibels; this is comparable to the noise level of a heavy diesel truck.

The threshold of hearing for a normal person is zero decibels, or a pressure equivalent of about 0.00006 lbs./sq.in. By way of comparison, the normal noise level of a business office is about 65 decibels, or a wave pressure of 0.107 lbs./sq.in.

Thus, in order for an air blast to damage a building, the associated noise level must be very, very high. Simply being able to hear an air blast at moderate sound levels is not evidence that the air blast was sufficient to cause damage.

If the air blast was not sufficient to break glass on the side of the building facing the point of origin of the explosion, then it is not possible that it could have caused any structural damage to the building.

If there is concern that damage could result from the concussion of a nearby blast, the sound levels can be electronically monitored, measured and recorded.

GROUND VIBRATIONS

When an explosive is detonated, it is generally intended that most of the explosive energy be directed into the surrounding ground, rather than into the air. The general purpose of using an explosive is to shatter rock or other impediments to construction.

Two kinds of vibrations are created when a blast is set off:

1. Surface or shear vibrations, and
2. Longitudinal vibration.

Roughly 95 percent of the blast energy that reaches a structure through the ground does so in the form of surface vibrations.

There are three common ways by which vibration levels can be measured: displacement, velocity, and acceleration.

Displacement, usually given in inches or millimeters, measures the total back-and-forth movement the soil particles make as the shock wave passes through them.

Velocity, usually measured in inches per second or millimeters per second, is a measure of the peak speed of the soil particles as the shock wave moves through.

Acceleration, measured in inches per second per second, or millimeters per second per second, is a measure of the change in soil-particle velocity over time as the shock wave passes through the particular location. Of the three methods, studies have found that velocity measurements are the most reliable for predicting structural damage.

In a ten-year study, the U.S. Bureau of Mines found that there was little chance of structural damage if the shock wave passing through a given area had a peak particle velocity of less than 2.0 in./sec. Minor damage could be expected at levels of 5.0 in./sec., and major damage could be expected at levels of 7.6 in./sec or more.

Interestingly enough, the threshold for human perception of blast ground vibrations in this study was found the be several times less than that for "safe" vibration levels for structures. In other words, a person will sense ground vibrations at much lower levels than are needed for damage to occur in a structure. In certain blast frequencies, the levels which a person may consider intolerable were found to be still less than levels which would cause actual damage to a structure.

The 2.0 in./sec. level has been adopted in several states as a legal criterion for "safe" blasting vibration levels for structures. Some states have opted for "safe" blasting levels of 0.5 in./sec., as was recommended in a follow-up Bureau of Mines study. This extremely conservative standard is now used to preclude even the most superficial damage to older, extremely delicate plastered residences, and to accommodate the lower threshold of human tolerance to such vibrations.

PREBLAST SURVEYS

In some communities, it is a legal requirement that a building or residence be surveyed prior to any nearby blasting. This usually involves making a photographic record of the building. If any claims of damage from the blasting occur, the claimed damage can the be compared to the photographic record, to examine whether any changes have occurred.

This system works well in averting a large number of unfounded claims. Often, building owners will not carefully inspect their walls and foundations for cracks until they actually feel the blast vibrations. Because human perception of the blast vibrations begins at levels far below that at which damage will occur, the owner's "sense" of the vibration levels is exaggerated. He is sure that some damage has occurred because he personally "felt" its magnitude. Thus, when cracks or fissures are found that "were not there before", it is common that the blame is laid on the blasting work.

BLAST MONITORING

Sometimes, it is necessary to monitor the vibrational levels of the blasting in "real time", that is, as it actually happens. This is accomplished through the use of an electronic seismograph monitor.

A seismograph monitor will measure and record ground vibrations. It can often be equipped to measure the air-blast shock wave as well. Normally, the seismograph is placed at a location near the building; best, of course, is placement between the building and the point of origin of the blast. This will cause the seismograph to intercept the ground (or air) vibrations just prior to the point where they reach the building. Seismographs are excellent insurance against unfounded claims.

POSTBLAST SURVEYS

As is often the case, many situations occur in which all of the data must be reconstructed after the damage has occurred. In these cases it is necessary to inspect the premises and review the blasting logs of the contractor.

A blasting log is a fundamental document which lists the location where the charge was set, the type of charges used, the total weight of the charge used, the charge weight per delay, the time of delay, the method of detonation, and the location of the nearest building or structure.

In most states, it is a legal requirement that the contractor keep and maintain such a log. Often, the failure to properly maintain a blasting log causes the contractor to practically forfeit any defense against blasting-damage claims.

With proper log information it is possible to calculate the level of vibration which the building would have experienced under the circumstances, and to determine whether the vibration levels would have been too high.

Normally, a "look-see" of the building is also needed, especially when the calculated vibration levels are higher than 0.5 in./sec. This is done to ensure that the building has no undue "delicateness" that might make it susceptible to damage at levels of 0.5 to 2.0 in./sec.

During the "look-see", the cracks and fissures can be examined to review whether they are consistent with blasting signature pattern damage. Simple items such as paint within a crack, dead insects or debris within a crack, old sealants and the like will tip off the inspector that the crack has been open for some time. Further, the analysis of the building structure for such factors as settlement or construction deficiencies will also provide information about the validity of the association between the particular cracking and the blasting.

HEAVY EQUIPMENT VIBRATIONS

Heavy equipment vibrations, such as jackhammers, headache balls, pile drivers and large trucks can cause vibrations similar to those of blasts. While the frequency distribution of the vibration levels may be different and the duration may be longer, the same basic data and equipment may be used to evaluate the possible deleterious effects.

It is often useful to monitor construction equipment vibration levels while actual work is proceeding on a spot basis, particularly if neighbor complaints are anticipated. This is usually done at various sites located between the work and the neighboring buildings.

In some cases, after the fact, it is possible to recreate the vibrational circumstances which are believed to have caused damage, and to then monitor them with a seismograph. Such evidence can be used to establish reasonable estimates of the vibrational levels at the time the work was actually being performed.

Case Study

In a Midwestern city, jackhammer work was done across the street from a local bar and grill. The pavement was broken up and a new pavement was laid. The bar and grill was in an older building with stucco exterior over brickwork.

Just as the street work was completed, the stucco on the front exterior wall of the bar and grill detached from the brickwork and fell off the building onto the sidewalk in front of the building. No other damage occurred. The owner of the building claimed that the vibrations of the jackhammers loosened the stucco and caused it to fall off the building.

Even though the street work was largely completed, it was decided to use the same jackhammers and break up a small portion of the new pavement. The work was to be performed as it had been done before. While this process was occurring, a seismograph was set up next to the location on the building where the stucco had detached.

The seismograph readings were taken as the simulated work progressed. The level of vibrations recorded was far below levels which could have caused any problems for the building.

Because of the low vibrational readings, a closer scrutiny was made of the stucco wall. It was found that the wall had sustained significant water seepage from the roof, and that this seepage was most severe in the area where the stucco had detached from the brick.

Observations of the brickwork found that none of the stucco had been left sticking to it; the stucco had come off in one piece. A check of the weather records for the night prior to the accident showed that it had rained.

Thus, it was concluded that the stucco had detached not because of the excessive vibration due to construction equipment, but because rainwater had leached away the bonding between the stucco and the brickwork.

REFERENCES

1. Bureau of Mines Bulletin No. 656, *Blasting Vibrations and Their Effects on Structures* .
2. Bureau of Mines Report of Investigation, 1980, *Structure Response and Damage Produced by Ground Vibration from Surface Mine Blasting* .
3. American Insurance Association, *Blasting Damage - a Guide for Adjusters and Engineers*.
4. Donald Dressler, P.E., *Damage Assessment to Structures*, University of Kansas Thesis, 1986.

CHAPTER 9

Explosions

INTRODUCTION

An explosion is a sudden, violent release of energy. It is usually accompanied by a loud noise, and an expanding pressure wave of gas which decreases in pressure with distance from the origin. Explosions resulting from the ignition of flammable materials may also be accompanied by a fireball containing high temperatures, which can ignite nearby combustibles.

Explosions caused by the ignition of flammable materials are classified into two main types: **deflagrating** and **detonating**. A deflagrating explosion is characterized by a relatively slow, progressive burn rate of the exploding material. A detonating explosion is characterized by a relatively quick burn rate, high energy release, and high peak explosion pressures. A general distinction between deflagrating and detonating explosions is that the former has subsonic flame propagation rates, while the latter have supersonic flame propagation rates.

Examples of deflagrating explosives include:

1. Explosive mixtures of natural gas and air at room conditions.

2. The decomposition of cellulose nitrate (an unstable compound often used in propellants).

3. Black powder.

4. Grain dust.

Examples of detonating explosives include:

1. Dynamite.

2. Nitroglycerine.

3. Mercury fulminate.

4. Trinitrotoluene (TNT).

5. Ammonium nitrate.

Under special conditions, a normally deflagrating explosive can be made to detonate. Such special conditions include the application of high pressures, strong sources of ignition, and long flame runup distances.

When an explosion occurs in an unconfined, open area, the pressure wave will harmlessly expand until the pressure becomes insignificant. When an explosion occurs in a confined space, the pressure wave will push against the confining structure.

When an explosion occurs within a typical building, the building is generally damaged. While most building codes require that buildings be able to withstand externally applied downward loads due to snow, rain, ice, and wind, they do not require that the buildings be able to withstand the outwardly directed loads generated by an explosion.

In buildings, most accidental explosions are typically caused by some of the following:

1. Ignition of natural gas leaks.

2. Ignition of vapors from improperly stored gasoline, cleaning solvents, copy-machine chemicals, or other such flamma ble materials.

3. Ignition of liquid propane (LP) vapors which have leaked.

4. Ignition of grain dust, coal dust, flour dust, and other types of dust from combustible materials.

5. Ignition of certain types of fine metal powders, such as aluminum or magnesium.

6. Ignition of atomized flammable or combustible liquids.

Inspection of the above list finds that most accidental explosions that occur in buildings are of the deflagrating type.

Explosion materials can be ignited in a number of ways. The most common source of ignition is an electric spark. The following is a partial list of common sources of electrical sparks:

1. Electric motors (commutator-slip ring sliding contact.)

2. Loose electric plugs in wall sockets. When the appliance is turned on, the plug prongs may arc to the contacts.

3. Static discharge due to frictional action between two electrically dissimilar materials. (The effect is called triboelectrification, and can occur between any combination of gas, liquid and solid. A particularly hazardous situation is the pumping of flammable liquid from one container to another, especially a plastic container.)

4. Relays and switches.

5. Lightning.

6. Electric insect killers.

7. Thermostats.

Explosions can also be touched off by some of the following ignition sources:

1. Pilot lights.

2. Smoking materials.

3. Sparks from the sliding contact of metals with an abrasive, such as a grinding wheel, acetylene gas igniter, flint and steel, etc.

4. Stove or furnace electric igniters.

5. Hot surfaces; e.g., electric heater elements.

6. Radiant energy. Some chemicals need only to be exposed to sunlight to explode.

7. Heat from chemical reactions between items being mixed.

8. Running automobiles or other internal combustion engines (especially in confined garages).

BASIC PARAMETERS

In both deflagrating and detonating explosions, the maximum pressure occurs when the explosion is wholly confined (constant volume process), and the explosive mixture is close to stoichiometric concentrations. The maximum pressure for many hydrocarbon-based deflagrating explosion mixtures will range between eight and nine times that of the ambient pressure. The typical maximum pressure for a gaseous detonating explosion will be nearly double that of a deflagrating explosion.

Thus, if the ambient pressure was 14.7 psi (1 standard atmosphere), the maximum pressure for a typical deflagrating explosion could range from 118 psi to 132 psi. A gaseous detonating explosion might reach a maximum pressure of about 265 psi.

The following graph in Figure 1 depicts the pressure rise (gauge) of a methane/air deflagrating explosion at stoichiometric mixture, 14.7 psi ambient pressure, and 77 degrees Fahrenheit, in a 12-ft. diameter sphere. It is notable that the maximum pressure occurs in less than one second.

In deflagrating explosions, there is normally a mixture range of fuel and air in which explosions can take place. For example, natural gas, or methane, will explode when its concentration is between 5 and 15 percent in air. If the concentration is less than 5 percent or more than 15 percent, no explosion will take place.

The following table lists some common deflagrating fuels and their explosive limits.

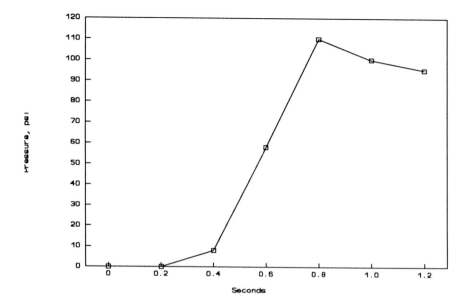

Figure 1. Methane Explosion: Pressure Rise versus Elapsed Time. Source: U.S. Dept. of the Interior, Bureau of Mines, Bulletin 680, p. 16.

Table 1. Explosive Limits -- Gases

Fuel	Limits (% v/v)
methane	5.0 - 15.0
ethane	3.0 - 12.4
propane	2.1 - 9.5
acetone	2.6 - 13.0
ammonia	15.0 - 28.0
gasoline	1.3 - 6.0
carbon monoxide	12.5 - 74.0
methanol	6.7 - 12.0

For dusts such as flour and grain, the lower explosive limits are about 0.02 ounces of suspended material per cubic foot of air. The upper explosive limits are not clearly defined.

The following table lists the lower explosive limits of some metallic dusts. Like the grain dusts, the upper explosive limits are not clearly defined.

Table 2. Explosive Limits - Metallic Powders

Material	Limit (oz/cu.ft)
aluminum	0.08
iron	0.12
magnesium	0.03
manganese	0.12
zinc	0.48

In Figure 1, the maximum pressure in a constant-volume methane deflagrating explosion was depicted. However, this was for the stoichiometric mixture, or the mixture where the gas and air volume leave no unburned fuel or unused oxygen. In general, as the amount of methane approaches the lower explosive limit, the peak explosion pressure rise will be less. Consequently, the resulting structural damage will also be less severe.

The following graph in Figure 2 shows the effect of varying the concentration upon the explosion pressure in a methane deflagration.

Likewise, the maximum flame temperature of methane in a deflagration explosion, 3905 degrees Fahrenheit, occurs at the stoichiometric mixture. As the lower explosive limit is approached, the maximum temperature will drop. The following graph in Figure 3 shows the relation of peak flame temperature versus the percent of methane in air, in a constant-volume deflagrating explosion.

The explosive limits of a fuel are useful in calculating the amount of fuel that was involved in an explosion. If the room or space in which the explosion occurred is known, its volume can be determined. If the type of gas that fueled the explosion is known, or assumed, then the amount of gas necessary to cause the explosion can be estimated from the explosive limits.

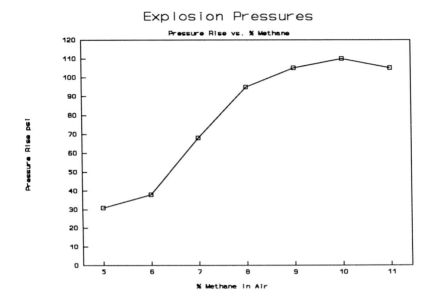

Figure 2. Explosion Pressures:
Pressure Rise versus percent Methane

Consider the following example of using the explosive limits to help diagnose an explosion: Suppose an explosion occurred in a small bedroom, 12' by 10' by 8', which contained a natural-gas space heater. The explosion occurred when a light switch, located halfway down the 8-ft. high wall, was switched after the room had been closed up for two hours. How much gas had leaked, at what rate had it leaked, and what size leak would be necessary to cause the problem?

Since natural gas is buoyant in air (sp. gr. = 0.55), much of the natural gas would collect on the ceiling and fill the room downward. Thus, the explosion would occur when the concentration at the light switch was at least 5 percent. (Light burns or singeing on the ceiling and walls would also have marked where the gas was present in explosive mixtures). Assuming a "cloud" or 5-percent methane/air in the top half of the room, then about 24 cubic feet of methane would have been needed. Since the room had been closed for two hours, the leak rate had to be 0.2 cu.ft./min.

Because most household gas systems operate at a pressure of about 0.5 psi, application of the Bernoulli equation finds that the

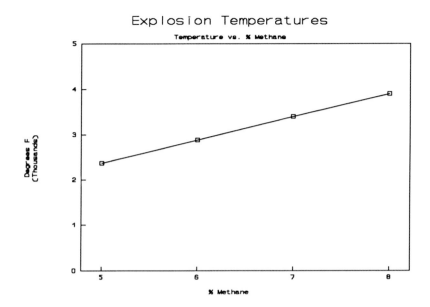

Figure 3. Explosion Temperatures:
Temperature versus percent Methane

escape velocity of the natural gas from the gas line or heater would be 2.32 ft/sec. Thus, to have leak rate of 0.2 cu.ft/min., a leak cross-sectional area of 0.21 sq.in. is required. This estimate of leak size gives a hint as to what to look for, then, in the gas lines and components around the gas heater.

DETERMINATION OF THE POINT OF IGNITION

In areas where there has been a buildup and subsequent ignition of flammable gases causing an explosion, the space where the gas accumulated will be evident from singeing and burns from the fireball or flame front. Wood items will be darkened, and thin, low ignition point items may be melted or burned. Due to the very short time available for heat transfer from the flame front to the item, items with higher thermal inertia will be less affected by the momentary flame front. As noted before, by determining the space where the flammable gases accumulated, an estimate can be made of the volume of explosive gas. However, within this space, or near to it, lies the point of ignition.

When a gas leak occurs, the gas will form an irregular fuel cloud that diffuses away from the leak. If the gas is methane and the air is more or less calm, the gas will slowly rise to the ceiling and accumulate in high areas. If the gas is propane, it will fall to the floor and accumulate in low areas.

As the fuel cloud expands and moves away from the point of leakage, it may come into contact with an energy source capable of igniting the fuel. Ignition will then occur when the fuel concentration in air exceeds the lower limit of explosion at that point.

In cases where the source of ignition is steady, like a pilot light, the cloud of gas will be ignited when the edge of the cloud reaches the pilot light. Where the source of ignition is intermittent, the cloud of gas may ignite after surrounding the ignition source.

The point of ignition may be determined from the damage vectors. During the explosion, a pressure wave and fireball radiates away from the point of ignition. The fireball will burn the side on which it impinges, but leave unburned the opposite side (the shadow side). Each item found burned or singed in this way will provide a single vector which points in the direction from which the fireball originated. Of course, the more severely burned items will have been closer to the ignition source than the less severely burned ones. The plot of all these vectors will converge at the point of ignition.

Similarly, the pressure rise associated with the explosion is likewise directed radially away from the point of ignition. The pressure wave will blow down walls, lift ceilings, blow out windows, dent appliances and ductwork, etc. The pressure wave will cause greater damage near the point of ignition than away from it. The plot of these vectors will also converge at the point of ignition.

It should be noted, however, that due to walls, hallways, and other features within a building, pressure waves and fireballs may not always travel outwardly like an expanding sphere for very long. Thus, the damage vectors may form a trail. That is why the relative intensity of the burn or pressure damage is noteworthy.

Thus, the preservation or documentation of the debris scatter after an explosion is important in order to determine the point of origin of

Figure 4.   Explosion which leveled a downtown Independence, Kansas building. No one was seriously hurt.

the explosion. Especially helpful are aerial photographs or sketches coupled with close examination of some of the components.

ENERGY CONSIDERATIONS

The amount of energy contained in an explosion is a direct function of the type of fuel, the amount of space in which it is confined, and the concentration of the fuel. In general, the energy of an explosion is dissipated in the following forms:

1. Acoustical: the blast sound

2. Kinetic: the displacement of objects away from the point of origin of the explosion.

3. Heat and expansion energy lost to the surroundings.

Because the pressure wave dissipates as the cube of the distance from the explosion volume envelope, if the pressure needed to lift a ceiling, push over a wall, or break glass in a window across the street can be determined, then the pressure of the explosion at the constant-volume envelope boundaries can be calculated. Knowing the room or space where the explosion occurred, it is possible to narrow down the list of possible fuels in an explosion. In short, it is possible to gauge the energy content of an explosion and deduce what kind of fuel caused it.

For example, it is known that single-pane common glass will break when a pressure wave of 0.5 lbs./sq.in. is applied to it. If the windows in a two-block radius have been broken out by an explosion, then the pressure level at the origin of the explosion can be calculated by a scaled inverse-cube relationship.

It should be noted that in an explosion, the constant-volume assumptions hold only as long as the structure stays intact. When the structure comes apart because it is too weak, and allows the pressure wave to escape, the process has become a constant-pressure process. Thus, most explosions are a two-step process: constant volume, and then constant pressure. If there has been severe structural damage to the building, generally the process is considered to be mostly constant-pressure. When the explosion has been largely contained and damage is slight, the process is mostly constant-volume. These considerations are important in deciding which thermodynamic relations apply.

REFERENCES

1. *A Pocket Guide to Arson Investigation*, Second Edition, Factory Mutual International, 1979, 25 pp. A quick and handy pocketbook of facts about fuels and investigative procedures.
2. *Fire Investigation Handbook*, U.S. Department of Commerce, National Bureau of Standards, NBS Handbook 134, August 1980, 187 pp. A general "how-to" handbook about fire investigations, which contains a basic descriptive chapter on explosions.
3. *Investigation of Fire and Explosion Accidents in the Chemical, Mining, and Fuel-Related Industries - a Manual* by Joseph Kuchta, U.S. Department of the Interior, Bureau of Mines, Bulletin 680, 1985, 84 pp. An excellent technical monograph about explosions, containing both theoretical and experimental information.

CHAPTER 10

Combustion

THE COMBUSTION PROCESS

Most forced-air furnaces used in homes and apartments use either natural gas or propane to provide heat. If natural gas is used, the gas is obtained from a local utility pipe system. If propane is used, the user must normally have a storage tank on site, which has to be periodically filled.

Natural gas is chiefly composed of methane. However, the natural gas that comes from the utility will also contain small amounts of ethane, carbon dioxide, mercaptan odorant and a number of other minor, basically inert gases.

The exact composition of utility-supplied natural gas will vary from city to city, and depends on the gas field from which it was taken. Thus the amount of heat obtained from a standard cubic foot of natural gas will also vary from place to place. Upon request, most utilities will provide heating and composition data. These parameters are routinely assayed by the utility for various contractual and regulatory reasons.

In general, it is typical for utility-supplied natural gas to contain between 85 and 95 percent methane.

Methane is a very simple organic gas. It is one carbon atom to which four hydrogen atoms are attached. When methane burns under ideal conditions, it produces heat and by-products of water vapor and carbon dioxide. A simplified chemical formula for this is given by

$$CH_4 + 2(O_2) \rightarrow 2(H_2O) + CO_2 + heat$$

or, methane + oxygen = water vapor + carbon dioxide + heat.

As a rule of thumb, a thousand cubic feet of natural gas (commonly called an MCF) will produce about one million BTU's of heat. Of course, the amount of heat actually available to the home or apartment depends on the efficiency of the furnace system.

ACID PRODUCTION

In more traditional forced-air gas furnaces, the flue gas is kept hot enough for the water to remain as a vapor until it exits the flue. This is because an acid is formed in the flue pipe if the water vapor becomes cool enough to condense into liquid. The temperature at which condensation occurs is called the **dew point**.

Condensing water and carbon dioxide gas will form carbonic acid, which is especially corrosive to sheet metal and other iron-based alloys. It will eat holes in the flue pipe. Because the acid solution is liquid, it can run down into other, more expensive areas of the furnace and similarly damage them.

This acid can be a safety problem, because it can allow flue gases to escape from the flue pipe into occupied areas. If the corrosion occurs in the heat exchange of the furnace, it can allow flue gases to enter the ventilation air stream to the residence interior. Such situations can cause loss of life due to carbon monoxide poisoning.

Corrosion of the flue pipe can also allow hot flue gases to escape and dry out nearby wood, paper or other flammable materials. In some cases, the loss of moisture can lower the ignition point of the nearby flammable material until it can be ignited by the hot flue gases themselves.

Generally, acid formation will occur when the flue pipe riser is excessively large with respect to the amount of flue gas it must carry. This allows cool air to be drawn in from the top in a convective cell. Acid formation can also occur if the flue pipe is too long, or passes through an unusually cool portion of the building.

In the newer, high-efficiency units, the over-all efficiency may exceed 90 percent, as compared to 60-70 percent in conventional furnaces. This means that 90 percent of the potential heat produced by burning will be actually delivered to the living space of the residence. However, the improvement in efficiency is accomplished by removing heat from the flue gases. This causes the water vapor to condense into liquid water.

The acid problem is handled by draining the liquid condensate from the unit by a plastic pipe. With the water vapor removed from the flue gas, the total amount of flue gas is reduced by two-thirds. With this reduction in flue gas volume, and the resulting reduction in flue gas temperature, a conventional flue pipe riser becomes unnecessary. Generally, a small plastic pipe is run to an outside wall to vent the remaining gases to the outdoors.

CARBON MONOXIDE

Carbon monoxide is an odorless, colorless gas. Unlike carbon dioxide, carbon monoxide is toxic. The usual industrial standard for allowable carbon monoxide is 35 parts per million.

The properties of carbon dioxide and carbon monoxide are often confused. Carbon dioxide is essentially an inert gas to humans; its inhalation does not do any good, but doesn't do a great deal of harm, either. Carbon dioxide occurs naturally in the air we all breathe. Breathing nothing but carbon dioxide will suffocate a person; breathing carbon dioxide along with the proper amount of oxygen is benign. This is why carbon dioxide is classified as an asphyxiant.

On the other hand, carbon monoxide is an active poison. Even if there is sufficient oxygen present, carbon monoxide will supersede oxygen uptake by the body. In fact, carbon monoxide has a greater affinity for blood than oxygen by two hundred times.

This means that even when there is the proper amount of oxygen present, the body will preferentially consume the poisonous carbon monoxide. Some people mistakenly believe that getting fresh air to a carbon monoxide poisoned person will quickly alleviate the problem; this is not the case. While it certainly is a good idea to provide fresh air, a person who has been poisoned by carbon monoxide should seek medical attention immediately. Fresh air will not take away the poison that has already entered the bloodstream.

With respect to forced-air furnaces, carbon monoxide is normally produced in small amounts by the combustion process. While ideally the products of combustion are carbon dioxide and water vapor, unfortunately, ideal burning conditions are not practically attainable.

Carbon monoxide is produced in the place of carbon dioxide when there is not enough oxygen available to match the fuel. This does not necessarily mean that carbon monoxide will not be produced if the correct mixture of air fuel is supplied for combustion: even when the correct air/fuel ratio is provided, some carbon monoxide will be produced.

Why this occurs can be explained by the following analogy. Imagine a game of musical chairs where there is a chair for each player. When the music stops (combustion occurs), some players may be bunched up in one area, while in another area they may be sparse. Thus, a player may still be left without a chair because when the music stopped, the available chair was across the room instead of nearby. In our game of musical combustion, the carbon atom has only a limited amount of time to find enough oxygen to form carbon dioxide. If it misses that chance, carbon monoxide is formed instead.

Fortunately, if the air/fuel ratio is properly "tuned", the amount of carbon monoxide produced will be very small.

However, carbon monoxide can be produced in larger amounts when a furnace is not properly maintained. If the air/fuel ratio is improperly adjusted by the installer or owner to be fuel-rich, the amount of carbon monoxide in the flue gas will be increased proportionately to the amount of excess fuel. This is why, from time to time, the air/fuel ratio of a furnace requires readjustment. Not only does this reduce the amount of carbon monoxide produced, but it makes the furnace operate more efficiently and saves the owner some cash.

The amount of fuel supplied by the burner orifice is proportional to the square of the orifice diameter. Thus, if the diameter of the orifice were doubled, the amount of gas supplied would be quadrupled.

DUAL FUEL FURNACES

Some furnaces can be used for either natural gas or propane. Such furnaces will have a conversion process to properly prepare the unit for whichever gas is used.

If a furnace is set for natural gas and is mistakenly connected to a propane-gas system, the air/fuel ratio will be far too fuel-rich. Propane is denser than natural gas for the same amount of heat and requires a smaller orifice. A natural-gas orifice would be too large.

In cases where this mixup occurs, not only will the amount of carbon monoxide be significantly increased, but a large amount of soot (unburned carbon) will be produced in the flue gas. This may be deposited in the flue-pipe riser, and may create a fire hazard similar to creosote deposits in a wood-stove flue.

OVERPRESSURIZATION

Not only is the amount of fuel supply proportional to the orifice size, but it is also proportional to variations in line pressure. A large increase in line pressure will force more fuel through the orifice and also cause the mixture to be too fuel-rich. Most natural-gas furnaces operate at pressures around 11 to 15 inches of water, or about one-half pound per square inch.

Occasionally, there are failures in the supply system that will overpressure the lines. These failures may include breakdown of pressure regulators or a mistaken hookup to a higher-pressure gas line. When this occurs, the burner flames will be very high, and the flame will appear yellow instead of the usual blue. Sooting will also likely occur. While this effect is occurring, popping noises may be heard from the furnace due to the overheating of metal parts. In some cases a "hiss" may be heard also; that would be the sound of the high-pressure gas passing through the orifice.

When overpressurization occurs, there is a significant risk of fire due to the overheating of the furnace. Nearby flammable materials may become sufficiently hot to ignite. This is especially true of furnaces housed in closets where the occupant has stored cleaning chemicals or other items near the furnace.

CROSSOVER

Normally, the air for combustion of the fuel is kept separate from the air that heats the space of the residence. Thus, even if the furnace is producing excessive carbon monoxide, theoretically the potential poison would be harmlessly vented to the outdoors.

Air for combustion will be drawn in from the room where the furnace is set, mixed with the fuel gases, burned in the combustion chamber, and exit the premises by means of the flue-gas riser. The air to be heated will be routed through the return air ductwork and blower fan, then around the combustion chamber through the heat exchanger, and will then be sent back to the living space areas. The air for combustion and the air for heating the residence are not mixed.

However, a hole in the heat exchanger will allow crossover to occur. Such a hole will allow flue gases to enter the air stream being sent to the living areas. A blockage in the flue can also cause crossover. In most flues, there is a small damper valve or pipe arrangement to allow air from the furnace room to be drawn into the flue. Flue gases can enter the residence space via this spot if the chimney is blocked.

Unfortunately, the same lack of maintenance that allows excessive carbon monoxide to be produced unchecked also allows for chimney blockages or holes in heat exchangers to occur. The combination is deadly.

CHAPTER 11

Determining The Point of Origin of a Fire

BACKGROUND

Fire losses in the U.S. are substantial, and are growing. In 1988, fire losses in the country totaled nearly ten billion dollars, averaging $39.12 for each person. The total amount of fire losses in the U.S. has increased 4.1 times between 1970 and 1988; during the same time period, the consumer price index rose by only 3 times. Thus, the increase in monetary losses caused by fire is outstripping inflation, and growing at a real-dollar rate of 2 percent per year.

In 1988, a total of 6,215 persons were killed in fires, and another 30,800 were injured. Of the total deaths, 66% occurred in one- and two-family dwellings. Of the total injuries, 56% occurred in one- and two-family dwellings. While there were 745,000 fires in all structures in 1988, one- and two-family dwellings accounted for 58% of all structure fires, and 35% of the total monetary losses by fire in all categories.

As indicated by these statistics, fires in one- and two-family dwellings constitute the largest single category of occurrence, monetary loss, and personal injury. For this reason, the following discussion concentrates on fires in residential-type structures.

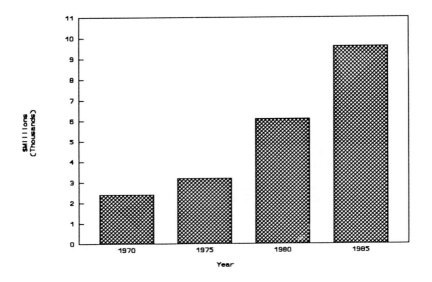

Figure 1. Total Monetary Losses by Fire in U.S.

FIRE

A fire is simply a type of chemical reaction: it is a rapid, self-sustaining oxidation reaction where heat, light and other by-products are produced. While it may seem almost a trivial point to state, in order for fire to take place, three things must be present: combustible fuel, air or another oxidizing material, and an energy source for ignition. This maxim forms a basis for assessing the point of origin of a fire: the point of origin must be where all these three components were present at the same time and place.

INITIAL RECONNOITER

In viewing a fire scene for the first time, it is often best not to rush directly into the fire-damaged areas in search of the origin. Most investigators will first reconnoiter the fire scene to observe which areas did or did not burn. This is important for two reasons: first and obvious is the fact that the areas which did not burn do not contain the point of origin. Often, knowing what was not burned by the fire allows the elimination of many theoretical point-of-origin possibilities. The second reason is to determine the extent of fire damage to the building and to examine it carefully for structural weakness before entering it. A crippled building can be a death trap.

THE LOW POINT

In general, heat can be transferred from one point to another by only three means: conduction, convection and radiation. Of these three means, fire spreads more often by conduction and convection, with convection being the most rapid mode of spread. Only in the most intense fires, such as high-energy fuel fires, does fire spread significantly by radiant heat transfer. Typically, this is not an important heat transfer mode in residential structure fires.

When heat transfer occurs by means of convection, the hot burning gases rise upward, contact nearby unburned material, heat it to ignition, and the process then repeats.

Heat transfer by conduction proceeds a little more slowly: Heat from a burning area is conducted directly through the material to immediately adjacent unburned material. Heat flows from the higher-temperature area (the area on fire) to the lower-temperature area (the area not on fire), causing its temperature to increase. When its temperature increases sufficiently to cause ignition, the process then repeats.

The higher rate of heat transfer by convection over conduction is the primary reason why the classic "V" burn pattern occurs. If a vertical panel of plywood or cardboard is ignited near its bottom, the fire will spread upward (primarily by means of convection) perhaps 2 to 10 times faster than it will spread laterally left and right (by means of conduction and some convection), and perhaps 10 - 100 times faster than it will spread downward (mostly via conduction). The resulting burn damage is in the shape of the letter V. The point of origin of such a fire is just above the notch of the V.

Thus, investigators often look for the lowest point of significant burn damage, and observe whether fire damage appears to fan out laterally and upward from that area. If the fire began on a wall or at the base of a wall, the damage pattern will appear like a "V". If the fire began in a cellular structure, like a pallet stack of small boxes or tightly spaced shelves of stock, the damage pattern will be like an irregular inverted cone, that is, like a three-dimensional "V".

FALL DOWN

Fall down is a general term used to indicate materials that have become displaced to new locations, generally to floor areas, because of fire damage. For example, in a particularly severe fire, roof shingles might be found in the basement area because the fire consumed the supporting structures between the two areas.

Fall down can create problems in determining the point of origin. Flaming fall down debris can produce secondary points of origin and "V" patterns which can be mistaken for the primary point of origin.

In general, such secondary points of origin can be distinguished from the primary point of origin by the following:

a. Since they occur later in the fire, the lateral spread from them is smaller.

b. Sometimes the piece of fall down which caused it is located nearby and is recognizable.

c. The fire at that point will have no other source of ignition except for the piece of fall down, which would normally not be associated with ignition. For example, it is not logical that a timber normally found in the attic could be the primary source of ignition for a fire which started in the basement.

CENTROID METHOD - UNWEIGHTED

Imagine that a large, square-shaped, sheet of paper, cardboard, or other plane-shaped combustible material is held so that its plane is horizontal. Imagine also that some point in the plane near its center is ignited. The resulting fire will burn away from the initial point of ignition more or less equally in all directions in the plane. Essentially, it will burn a big, circular hole in the item. At the center of that burned hole is the point of origin of that fire.

The above is an example of the underlying principle of the centroid method. This method is particularly useful in single-story structure fires, where the building is made of more or less the same materials and same type of construction throughout. This homogeneity of materials and construction means that lateral fire spread rates are similar in all directions.

Basically, the method works as discussed in the above example. The extent of burn damage in the structure is noted; then, the center of burn damage is determined. This can be done by "eyeball", or by more sophisticated methods including graphic integration.

Graphic integration involves the use of a drafting instrument called a planimeter, along with a detailed sketch of the fire-damaged areas. The planimeter is used to measure the total plane area encompassed by fire damage. The position of the centroid, or center point of the fire-damaged area, is calculated using the algebraic equations for plane centroids. The origin of the fire is in the centroid of the fire-damaged area.

Some discretion must be used in applying the centroid method, to allow for fires that start near "an edge" or "corner". In the previous example using paper, the fire had plenty of material to burn in any direction of travel in the plane, and was put out before it reached an edge. However, if the fire were begun at a corner or edge, it would burn away from that point in a skew symmetric fashion. For example, a fire initiated at an edge would burn away from the edge in a semi-circle pattern. A fire initiated at a corner would burn away from the corner in a quarter-circle pattern.

In short, the symmetry produced by more or less equilateral fire travel rate is affected when the fire must stop at a boundary for lack of fuel.

This shortcoming can be overcome by the use of mirror symmetry. For example, suppose that a fire occurred in the center of the east wall of a square-shaped building, and was put out before it spread more than 50% to the west side. The fire pattern would then appear half-moon-shaped in the plan view. By placing a mirror alongside the east edge of the building plan, the burned area reflects to the mirror and a round burn pattern is observed. The centroid method is then applied to both halves, the real and the mirror image, to find the point of origin. For a corner fire, two mirrors (or two mathematical "reflections") are needed.

Because the algebraic equations will produce an "exact" answer for the location of the centroid, it might be assumed that the centroid method produces an "exact" solution for the point of origin. This is not true, except for the most homogeneous of situations. Small variations in lateral flame spread rates affect the accuracy of the algebraic solution. However, the centroid method does generally find a centroid very near the actual point of origin, and the method can be easily adapted for computer use.

There are inexpensive CADD (computer assisted drafting and design) programs available that will allow the building plan to be laid out on a computer screen, and the fire-damaged areas to be drawn in . Some of the programs will then automatically calculate the area of fire damage and even automatically find its centroid. It is also possible, using inexpensive computer-store hardware and software, to scan actual photographs of the fire scene and work directly from the photographs. This, of course, requires that the photographs be taken from appropriate vantage points.

CENTROID METHOD - WEIGHTED

Similar to the first centroid method, this method assumes that the point which has been subjected to the longest duration of fire, and has consequently sustained the severest damage, is the point of origin. As before, it also assumes a generally homogeneous structure in terms of materials and construction. This method normally entails simply observing where the severest damage in the structure is located in terms of fire consumption of materials. An area whose materials have been fully consumed by fire is nearer the point of origin than an area only partially consumed.

Generally, this method is applied by "eyeball" by the fire investigator. However, the method does also lend itself to more sophisticated computational analysis in the following way:

First, a sketch drawn to the scale of the building is made, and then using a numerical rating system, fire-damaged areas of the building are rated. Using a ten-point system, an area with no fire damage would be rated "0", and an area fully consumed is rated "10". An area partially consumed is rated between 1 and 9, depending on the severity of fire damage. Thus the sketch will eventually appear like an unpainted "paint-by-number" picture, or like a contour map with the elevations posted between contour lines.

Figure 2. Fire damaged chair. Origin of fire was at observer's right side; note there is little damage below cushion area.

The centroid of the fire-damaged area is then calculated using the rating system to "weight" the areas. The previously discussed centroid of fire damage method assumed that all fire-damaged areas were equal; in this method, by contrast, the fire-damaged areas are distinguished by severity, and the more severely burned areas count more than less severely burned areas.

One note of caution should be observed when using this method: Not all homes burn homogeneously. For example, if a fire had started in a kitchen, and then spread to a storage room where an open 55-gallon drum of flammable liquid was stored, the most severely damaged fire area would be the storage area, not the kitchen. Thus, the initial reconnoiter of the damage should note where there were fuel concentrations that might skew the weighting.

FIRE SPREAD INDICATORS - SEQUENTIAL ANALYSIS

The point of origin of a fire can also be found by simply following the trail of damage from where it ended, backwards to where it began. Where several such trails converge, that is the point of origin. In essence, this method involves determining what burned last, what burned next to last, what burned before that, etc., until the first thing that burned is found.

In the same way that a hunting guide interprets signs and markers to follow the trail of game, a fire investigator looks for signs and markers which may lead to the point of origin. For example, a fast, very hot burn will produce a shiny type of wood charring with large alligatoring. A cooler, slower fire will produce alligatoring with smaller spacing and a duller-appearing char. Fire breakthrough or breaching of a one-hour rated firewall will likely take longer to occur than fire breakthrough of a stud and wood-paneled wall.

The interpretations of many such markers, "V" patterns, etc. are then combined into a logical construct of the fire path. So that a "false" trail due to fall down is not followed, several such fire paths are analyzed, beginning at various end points of the fire to ensure that they all converge at the same point of origin.

The advantage of this method is that no special assumptions need be made concerning structural homogeneity. The disadvantage of this method is twofold: First, that it relies on the individual skill and knowledge of the investigator to find and properly interpret the markers; second, that it assumes that all the markers necessary to posit a logical construct of the fire sequence are present and can be found. This is not always true. Sometimes the severity of the fire or the fire fighting activities themselves destroy significant markers, or simply cause them to be lost in the jumble of debris. The match that starts a forest fire often is obliterated by the same fire - never to be found.

IGNITION ENERGY SOURCES

Since air is more or less ubiquitous in a one- or two-family dwelling, and flammable materials are also readily available, the only missing component generally needed to start a fire is an energy source for ignition. Often, the point of origin of a fire can be easily spotted by examining the potential sources of ignition energy. Thus, an inventory and examination of such energy sources in the fire damage centroid area is important in establishing a specific point of origin.

Such energy sources commonly include pilot lights, space heaters, electrical appliances, electrical cords and outlets, electrical wall switches, fluorescent light ballasts, fireplaces and chimneys, smoking materials, cooking equipment, lamps and heat lamps, and engine manifolds and exhaust systems.

Occasionally, the ignition of combustibles can be spontaneous, but this usually involves organic oils or materials that can readily undergo decomposition. It is commonly assumed that a pile of oily rags can catch on fire by itself. This is only true if the oil comes from an organic source (e.g. linseed oil, spike oil, or tung oil) and contains large amounts of organic acids which can react with air at room temperature. Stable oils, such as petroleum-derived motor oils and lubricants, generally do not ignite spontaneously.

Decaying fecal material covered with straw and without ready access to the air can also spontaneously ignite. However, this is a multi-step process involving bacterial decay, and typically occurs in barns or animal pens. The occurrence of such spontaneous combustion in one- or two-family dwellings is relatively rare compared to other causes.

COMBINATION OF METHODS

Few fires lend themselves to complete analysis by only one of the methods described. Most require a combination of methodologies. For example, it is common to determine a general area where the fire began, using one of the centroid methods, and then determine a specific point of origin using a combination of fire spread indicators and an examination of the available potential ignition energy sources.

Figure 3. TOP - burn pattern on vehicle hood; note location. BOTTOM - inside, burned fuel line was located under burn pattern on hood.

ARSON

Of the 745,000 structural fires which occurred in 1988, some 99,500 were considered to have been set deliberately, or were suspected of having been set deliberately. This is an arson rate of 13.4% . In other words, about 1 in 7 structural fires in 1988 was a suspected case of arson. Thus, most fire investigators are regularly confronted with arson when doing point-of-origin analyses.

In general, an arson fire is distinguished from a "normal" fire by the fact that it does not follow the rules for a normal fire. It will often be an "illogical" fire. Instead of a single point of origin, it may have

Figure 4. Despite formidable appearance, the origin of fire was traced to an LP gas leak using the centroid method.

several. They may be connected by a narrow burn pattern, reminiscent of a trail of poured liquid, or they may not be connected to each available energy source for ignition, or it may be at a place that normally has no combustibles.

 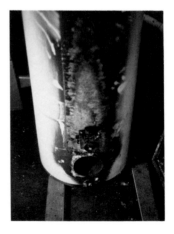

Figure 5. The "V" pattern on hot water tank, originating at gas control valve, left; at right, with gas control valve removed.

Many modern carpets, furniture materials, and textiles cannot sustain a flame. They will smoke profusely and perhaps burn when exposed to an external flame, but when the flame is removed they will self-extinguish.   A point of origin in such a material is a strong indicator of arson.

Another indicator of arson is a very hot and fast-moving fire in an area where the fire load (the items that will burn) would be expected to produce relatively low temperatures and a slow-moving fire. Unusually severe fire damage patterns on a floor that appear to outline a pool of ignited combustible liquid also suggest arson.

Unsecured areas accessible to transients and juvenile gangs are also popular places for deliberately set fires. And of course, some fires are set to mask burglaries or other crimes.

REFERENCES

1. U.S. Department of Commerce, *Fire Investigation Handbook*, National Bureau of Standards, NBS Handbook 134, August 1980.
2. Factory Mutual System, *A Pocket Guide to Arson Investigation*, 1979.
3. *The World Almanac and Book of Facts - 1991.*

CHAPTER 12

Arson
for Profit

THE BASIC PROBLEM
FACING THE ARSONIST

In order to be successful in his work, an arsonist must burn enough of the house, building or inventory for it to be considered a total loss. In that way, he will collect the money instead of having the insurance company rebuild the building or replace the items.

Thus, the fundamental problem facing the arsonist is how to set a fire such that it will have enough time to burn and consume the building before being spotted and put out by the fire department.

In cities where the response time of the fire department is just a few minutes, this means that the arsonist will likely need an accelerant to speed the fire along, and it means that he will likely have to set the fire in several places more or less simultaneously. He may also choose to set fires in strategic areas, such as stairways, to speed the spread of the fire to structurally important areas.

A further complication to the arsonist's problem is that he must set the fire, remain unseen, and avoid hurting himself. It is not uncommon for less astute arsonists to burn themselves as they attempt to ignite liquid gasoline on the floor in a room full of explosive gasoline vapors.

In some cases, the arsonist will actually be part of a team.   The person who will collect the insurance hires a second person to actually set the fire and assume the risk of getting caught.   This allows the insured to be seen and have an alibi at the time of the fire.   Of course, there is an inherent problem in this arrangement: in a pinch, will both partners "keep mum"?

USUAL PHYSICAL CHARACTERISTICS OF AN ARSON FIRE

The ways in which an arsonists solves his "problems" are the same ways that provide identification of an arson fire.   The following is a short list of the more common characteristics of an arson fire:

1. Multiple origins of fire, especially several points of fire origin that are unconnected to one another; that is, there is no trail of fire between them to indicate that one set the other.

2. Fire originates in areas where there is no rational ignition potential, for example, fire origination in a fireproof carpet.

3. Use of accelerants such as gasoline, kerosene, turpentine, etc.   Often these are noticeable by their odor, the "pour" patterns they produce on floors, and by chemical analysis. Many times two point of fire origin will be connected by an accelerant "pour" pattern.

4. Building furniture or items will have been abnormally rearranged to speed up the fire.

5. Buildings or residences that are "missing" personal items that normally would be present.   These items might include family photograph albums, special collections or instruments, trophies, prized dresses, tools, etc.

6. Buildings or residences with "extra" items that appear out of context.   This is done to "beef up" the amount of contents lost in the fire.

7. An unusually fast-consuming fire for the time involved, and a very high-burning temperature in areas where the fire load is quite ordinary.

8. Tampering with fire protection systems such as sprinkler systems, detection systems etc.

SECONDARY CHARACTERISTICS

In addition to the previous physical characteristics, there are other secondary characteristics that may suggest there has been arson. For example, some arsonists may call in a number of false alarms, or set some small fires in the neighborhood to actually measure the response time of the local fire department. In this way, they determine the time they have to work with. Thus, an unusual number of recent false calls or a number of small fires set near the building may indicate someone was timing the response of the fire trucks.

Sometimes the value of a building can be artificially jacked up by reselling it several times among acquaintances. For example, Joe buys a decrepit building for $40,000 and sells it to his brother Frank for $50,000, although no money actually changes hands. Frank then sells it to his cousin Thelma for $65,000, and Thelma sells it to Joe's wife for $80,000. At each point in the various transactions the building is insured for the sale price.

At some point the building burns down, and the last "owner" collects the inflated sale price insurance money on it.

Thus, if there has been an unusual number of buying and selling transactions on a building that has marginal value preceding a fire, it may indicate arson was planned well in advance.

Of course, very few cases of arson for profit occur in buildings that are underinsured.

ARSONIST PROFILE

On the average, there are some 18,000 arrests a year for arson. This is about 30 percent more than the total arrests for embezzlement, and is just slightly less (7 percent) than the total arrests for murder and non-negligent manslaughter.

The average arsonist is male. Women arsonists constitute only some 11 percent of the total arrests. Most arsonists are over 18 years of age; in fact, 52 percent of the total number arrested for arson are over 18; only 42 percent of the men arrested are under 18, and only 30 percent of the women are under 18 years old.

CHEMICAL ANALYSIS

When an accelerant has been used, it can often be detected visually by the "pour" pattern that occurs on the floor where it has burned.   A "pour" pattern is the damage made where the liquid accelerant has burned the floor where it was in contact with it.

In the case of wood floors, the burn damage will be a well-defined char pattern. This char pattern will often appear to "drain" along the drain gradient of the floor. In the case of concrete floors, the burn damage will be spalling in the areas where the accelerant contacted the floor.

In the "pour" areas, some of the accelerant will soak into the floor materials, thus allowing for chemical detection and identification. However, to stand up in court later, this must be done properly. One common mistake in taking samples for analysis is to use plastic sandwich bags or something similar; such bags themselves can contaminate the accelerant sample and render it useless.

One good way to obtain samples is in metal cans with lids that can be hammered down for a seal, such as unused paint cans. Such containers will also preserve vapors that may come off the sample. By using head space analysis, everything that was put into the can, even the vapors, can be checked for of possible accelerants.

One less desirable method of detecting accelerants is the solvent process. In this process, the sample is first washed with a solvent, often an alcohol. The idea is that if there is an accelerant present in the sample, such as gasoline or kerosene, the solvent will "pick up" some of it during the wash. The solvent is then analyzed in a gas chromatograph (GC). The problem with this method is that if the accelerant was chemically similar to the wash solvent, it can easily be missed.

An excellent method of detection is the head-space analysis method. In this method, the metal can itself is gently heated, and any accelerants are vaporized into the "head space". By their nature, accelerants are light distillates that vaporize easily. The vapors are then run through a GC machine for analysis; there is no wash to get in the way.

It is important to obtain samples as soon as possible after a fire. Delays in obtaining samples greatly diminish the chance of detection. After a week, it is doubtful that any of the typical accelerants can be identified. Because most accelerants are light distillates, they vaporize and diffuse away quickly.

Also, analysis for accelerants should be open to "creative" items. While gasoline and kerosene seem to be the accelerants of choice, other items such as linseed oil, cleaning fluids, mimeograph fluids, etc. have been used.

REFERENCES

1. U.S. Department of Commerce, *Fire Investigation Handbook*, National Bureau of Standards, NBS Handbook 134, August, 1980.
2. Factory Mutual System, *A Pocket Guide to Arson Investigation*, 1979.

CHAPTER 13

Electrical Shorting

PRIMARY VS. SECONDARY SHORTING

A large number of fires each year are attributed to shorting in electrical appliances or building wiring. Occasionally, some fires are conveniently blamed on electrical shorting simply because no other reasonable cause can be found.

Most fires in buildings that have electricity will eventually cause something to short out, which provides the needed "evidence" of shorting. Thus, it is necessary to discriminate between primary shorting and secondary shorting. The former is what initiates fires; the latter is itself caused by fire damage.

Primary shorting has the following general characteristics:

1. It occurs at the point of origin of the fire.

2. Heat damage to the conductor is more severe at the interior than at the exterior.

3. Significant movement or travel of the short has occurred.

On the other hand, secondary shorting has the following general characteristics:

1. It occurs in locations away from the point of origin of the fire.

2. The conductor interior may not be as severely damaged as its exterior.

3. Little movement or travel of the short has occurred.

SHORT CIRCUITS

In an electrical circuit, conductor pathways are provided so that electrical energy can "flow" to an appliance or component to perform some function. When a short circuit occurs, the electrical current flows through an unintended pathway or shortcut between conductors.

For example, in a toaster it is intended for the electrical current to flow through the cord and through the heating elements (the resistance load). If the cord becomes worn and shorting occurs at some point within the cord, electricity will "leak" across the cord without having to go through the main resistive load, i.e. the heating element of the toaster.

When the example toaster was in good working order (no shorts), the impressed alternating-current (ac) voltage caused current to flow through the heater, or resistive load. The equation for this is as follows:

$$V \sin w = I \sin (w+a) R_L \qquad (1)$$

where I sin (w+a) = the flow of current with time,
V sin (w) = the impressed ac voltage
and R_L = the resistive load.

The above equation can be simplified if the "effective" values (root-mean-square values) are used instead of the exact values. This gives the more familiar version of Ohm's Law:

$$V = I R_L \qquad (2)$$

To determine how much current is flowing through the resistive load, one need only to rearrange terms as follows:

$$I = \frac{V}{R_L} \tag{3}$$

The amount of electrical power consumed by the appliance is then

$$P = VI = I^2 R_L = \frac{V^2}{R_L} \tag{4}$$

For example, an ac voltage of 120 volts impressed on a toaster with a resistive load of 20 ohms will produce current flow of 6 amperes, and the appliance will use 720 watts of electrical power.

However, when a short occurs, there are now two resistive loads: the intended one (the heating element), and the resistance across the short. The resistance of the short and the resistance of the heater are in parallel with one another. In this situation, the effective resistance of both together is as follows:

$$R_{eq} = \frac{(R_S \times R_L)}{(R_S + R_L)} \tag{5}$$

Thus, if the short had a resistance of 5,000 ohms, the net effective resistance with the 20-ohm toaster would be 19.92 ohms. This would cause the total current to increase to 6.024 amperes, and the total power consumed to be 723 watts. While the heating element still only had a current of 6 amperes flowing through it, the short had an additional current of 0.024 amperes.

If the area around the short deteriorates and the resistance is reduced to only 1,000 ohms, then the net effective resistance will become 19.6 ohms, the total current will rise to 6.12 amperes, and the power consumed will increase to 734 watts.

From this example, it is seen that as the short deteriorates (the insulation between the conductors is reduced) and the resistance across the short drops, the appliance will take more power. Where does the additional power go? It produces heat in the area of the short.

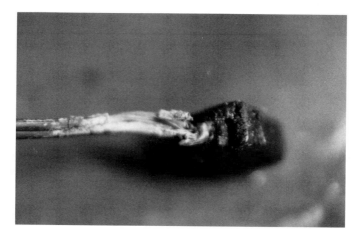

Figure 1. Overloading of electrical connector. Note how one conductor is more damaged than the other.

WHEN THE SWITCH IS OFF

The formula for net effective resistance of a short in parallel with the appliance load provides an opportunity to assess what happens when a person turns off the switch on an appliance that has a "hidden" short.

When the toaster in our example is turned off, no current flows through its heating element. However, the operation of the switch on the toaster does not affect the current through the short. If this is 1,000 ohms, it will still consume 0.12 amperes and 14.4 watts of power. It will do this day and night, without stopping.

Now, the power being consumed is converted into heat and is concentrated within a small area around the short. This heat causes more insulation to melt away from around the conductors. With either more exposed conductor area, or less insulation thickness, the resistance at the short can be reduced further. This is why:

1. Appliances tend to short out and catch fire when no one is around.

2. Appliances can still short out and catch fire even if they have been turned off. This is the underlying reason why

people are advised to unplug unused appliances when they are going away from their home for a time.

Parenthetically, it should be noted that approved modern cord insulation materials do not burn or catch fire like some older materials. They will smoke, melt and char, but will not sustain a flame. The plastics used in modern cord insulation are formulated so that they cannot burn with an open flame unless there is more oxygen than the atmospheric mixture provides.

For a fire to occur in a modern type of cord, melted insulation or conductor must fall on another material that can be ignited by the hot plastic, or the cord itself must be hot and placed tightly against something that can be easily ignited.

Thus, if the appliance is new and the diagnosis is that the cord shorted and the cord insulation caught fire, the latter is likely to be an erroneous conclusion.

BEADING

One of the traditional methods of determining whether shorting has occurred, is the observation of "**beading**". Beading is simply melted conductor that has formed droplets or beads near or at the end of the conductor, usually at the point where shorting has taken place and the conductor has separated.

The two materials most often used for conductors are copper and aluminum. The melting point of copper is about 1,892 degrees Fahrenheit, and its boiling point is about 4,172 degrees Fahrenheit. For aluminum, the melting point is 1,220 degrees Fahrenheit and the boiling point is 3,733 degrees Fahrenheit.

In "normal" building fires, the temperatures of burning wood, cloth, paper, etc., are not hot enough to melt copper conductors. Thus, the finding of melted copper is a strong indication of shorting. Shorting can cause copper to melt; however, it should be noted that some accelerants can also cause copper to melt.

However, the temperatures encountered in a "normal" building fire can be enough of themselves to melt aluminum. Thus, the finding of melted or beaded aluminum conductors should not be an automatic indicator of electric shorting.

Figure 2. Shorting of electric cord on edge of hole at rear of electric clothes dryer.

The amount of beading can often be indicative of how long the short was in progress. This is because the heat input to the short is often a function of the fuse rating, and it requires a specific amount of heat to raise the temperature of a conductor to melting point.

For example, consider a No. 14 copper conductor. Such a wire has a resistance of 2.525 ohms per 1,000 ft. at 68 degrees Fahrenheit. It weighs 12.43 pounds per 1,000 ft. Copper has a heat capacity of 0.0931 BTU lb. per degree Fahrenheit, and a heat of fusion of 75.6 BTU/lb.

If the conductor in question were 60 feet long, from fuse to short, and the fuse was rated at 15 amperes, and the resistance at the short was only 3.66 ohms, then the following would be true:

1. Total resistance at room temperature was 1.303 ohms.

2. Near the melting temperature, the total resistance was 5.34 ohms (resistance increases with temperature).

3. The amount of heat necessary to raise the temperature of the conductor to melting point was 268 BTU.

4. Assuming a power rate of 120 v.a.c. at 15 amperes, the electrical input to the short would have been no more than 1,800 watts, or 6,143 BTU/hr. To heat the conductor to near melting point would have required about 160 seconds.

5. After that point, the conductor would melt at a rate of 6.6 inches per second.

Thus, as is evident in the hypothetical example given above, the conductor will heat up to the melting point very fast, and the short will "travel" (i.e. melt one conductor wire) at a rate of 6.6 inches per second. Fire might occur as a result of melted copper falling onto something flammable, or hot conductor being in contact with flammable material.

FUSING

It is a mistaken notion that fuses or circuit breakers will wholly protect a building against fire by shorting. Fuses or breakers will trip only when their amperage rating is exceeded. Thus, as long as the short does not consume more than the amperage rating of the circuit, the fuse or breaker will not trip.

An oversized fuse is a safety hazard. While it allows a person to avoid the aggravation of changing fuses or resetting breakers in an overloaded system, it allows a short circuit to proceed unchecked to failure. Fuses should be sized so that they will blow when sufficient current flows to raise the temperature of the conductor in the circuit.

Figure 3. Power wire for decorative running-board lights on van. Wire shorted on sharp edge of sheet metal.

CHAPTER 14

Hail
Damage

HOW HAIL FORMS

Scientifically, hail is a form of **hydrometeor**. A hydrometeor is basically any type of condensation or sublimation of atmospheric water vapor. Hail is not very different from falling ice crystals, ice pellets or snow.

Hail occurs in association with strong convective cloud systems. Generally, the formation of cumulonimbus clouds is an excellent indication of possible hail fall. A cumulonimbus cloud is the classic towering thunder cloud: it has low, middle and upper altitude formations. The low portion may form as low as 500 feet above ground; this formation appears billowy, dense, tall, and may have cloud colors ranging from white to black. The top of the cloud may spread and flatten out, giving rise to the descriptions of thunderhead or anvil cloud.

When one views a cross section of a hailstone, various concentric layers of ice are observed, somewhat like an onion. This interior structure is used to support the Theory of Multiple Incursions with respect to hail formation. In this theory, hailstones are assumed to alternately ascend and descend through freezing and nonfreezing zones of the cloud formation; thus, the formation of hail requires a tall cloud formation, and abundant liquid moisture content.

The specific nature of hail formation means, essentially, that hail occurs almost exclusively in association with thunderstorms or thunderstorm weather systems. Thus, one of the first checks a person should do to verify roof damage by hail is to compare the date and time of the reported occurrence with local weather records.

SIZE RANGE

Hail ranges in size from less than that of a pea to several inches in diameter. The largest hailstone recorded in the United States was 17.5 inches in circumference, and was found on September 3, 1979 in Coffeyville, Kansas. The previous record holder had a circumference of 17 inches and weighed 1.5 pounds.

Stories of hailstones as large as bowling balls are in the same league with jackalopes and other Western extravagances. While hailstones the size of soft-balls (four inches in diameter) are possible, they are rare. Most hailstones by far are less than an inch in diameter.

HAILSTORM FREQUENCY

Since hail occurs in conjunction with thunderstorm activity, it follows that most hail damage will come in months with the greatest number of thunderstorms. As noted in the following table, the month with the greatest number of thunderstorms in the metropolitan Kansas City area is June. In fact, the months May, June and July typically have nearly half of the total annual thunderstorms for the area. Thus it is expected that these three months would have about half the annual hail-damage occurrences.

Table 1. Normal Thunderstorm Activity
in Kansas City Area

Month	Number of Thunderstorms
January	0.1
February	0.6
March	2.3
April	5.3
May	8.8
June	9.9
July	7.3
August	7.7
September	5.8
October	3.4
November	1.3
December	0.5

In a typical year there are 53 thunderstorms; nearly one a week. Hail falls only in only a fraction of the thunderstorms. The table on the following page shows the actual reported number of thunderstorms and hailstorms for the Kansas City area in 1987.

As shown, the number of thunderstorms in 1987 was 51 and the number of hailstorms was only 1. In 1986, in the same area, the numbers were 59 thunderstorms and 3 hailstorms.

Table 2. Hailstorm Activity in Kansas City Area

Month	Number of Thunderstorms	Number of Hailstorms
January	0	0
February	0	0
March	2	0
April	4	0
May	10	1
June	13	0
July	7	0
August	7	0
September	4	0
October	3	0
November	1	0
December	0	0

It might be reasoned that if one year had only one hailstorm, and another year had three, then claims of roof damage might be about one third less in the former than in the latter, and this might be valid if all claims were truthful and accurate, and all hailstorms had the same intensity. However, it is not unusual for a similar number of claims to occur in both years despite the difference in hailstorm frequency. This is because there are differences not only in hailstorm intensity, but also because some roof damages attributed to hail are actually due to wear and tear.

It would be a useful report for insurers to plot the number of hail-storms against the number of roof damage claims over several years. It might give an indication of whether or not such roof damage claims were being accurately assessed.

Figure 1. Hail "skid mark" and indentation on exterior wall of house.

HOW HAIL CAUSES DAMAGE

The severity of hail damage depends on two primary factors: how much kinetic energy the hailstone possesses, and how much of it is transferred to the roof in the impact.

Moving objects, such as hailstones, have an amount of kinetic energy that is a function of their mass and of their velocity. Algebraically this is expressed as follows:

$$KE = (1/2)mv^2 \tag{1}$$

In general, the shape of a hailstone is spherical. Since a hailstone is composed of ice, its mass can be calculated from its diameter by applying the formula for the volume of a sphere and multiplying it by the density of water (1 gram/cubic centimeter).

From experimentation, it is known that when a hailstone strikes a flat surface, it contacts an area about half the diameter of the hailstone. Further, the measure of hailstone impact is the amount of kinetic energy imparted to the impact area.

Hailstones usually impact a surface at an angle. The velocity component directed towards the surface is the cosine of the angle of approach from the perpendicular, multiplied by the velocity.

Combining all of the above into a single algebraic expression gives the following:

$$K\frac{E}{Area} = (\frac{4}{3})(p)(d)(v)^2(\cos^2 a)$$
(2)

where p = density of ice,
d = diameter of hailstone,
a = angle of approach from perpendicular,
and v = hailstone velocity.

Note: a = 0 degrees when the hailstone strikes the surface head-on; a = 90 degrees when the hailstone path is parallel to the surface.

It can be seen that as the angle of impact deviates from the perpendicular position, the impact severity diminishes. For example, at 30 degrees from perpendicular, the impact severity has been reduced by 25 percent; at 45 degrees, by 50 percent, and at 60 degrees by 75 percent.

From equation (2) it can be seen that the impact energy of a hailstone increases linearly with the diameter and also with the square of the velocity.

One additional point should be made: for roof damage to occur, there must be a transfer of kinetic energy from the hailstone to the roof. If a hailstone bounces off a roof with little change in absolute velocity, but only a change in direction, then very little energy has been transferred to the roof. If the hailstone loses most of its velocity at impact and rebounds feebly, then most of its kinetic energy has been transferred to the roof.

A measure of this energy transfer is called the **coefficient of restitution**, usually labelled "**e**". The coefficient of restitution is defined as

$$e = \frac{v_{(b)}}{v_{(a)}}$$
(3)

where $v_{(a)}$ = velocity before impact and
$v_{(b)}$ = velocity after impact.

The amount of energy per contact area transferred during an impact is then:

$$Q = (\frac{4}{3})(p)(d)(V^2)(1-e)^2(\cos^2 a) \qquad (4)$$

A ball bearing dropped onto a solid cement floor will generally bounce back almost to its original height. It has an "e" value of just under 1.0. The same ball bearing dropped into mud will likely rebound very little; it may have an "e" value of nearly zero.

WHAT ALL THIS MEANS

The preceding may appear a little complicated; however, the lessons to be derived can be summarized as follows:

Studies by the U.S. National Bureau of Standards have shown that damage to a roof in reasonably good condition will not occur until an impact energy per contact area of 4,000 g-m^2/cm-sec^2 is exceeded. In general, this means that with typical winds, it takes a hailstone size of about 2 inches in diameter to cause damage.

How would a person find out how big the hailstones were? One way would be to examine the impact marks on sheet metal surfaces and measure their diameters, remembering that the diameter of the mark is about half that of the hailstone.

Roofs that have little "give" are less likely to be damaged than roofs that have a lot of bounce or "give" to them. This is because a stiff or firm roof has a high value of "e".

Built-up roofs (BUR's) that have an adequate gravel cover will not be damaged by hail until the hail diameter is quite large. This is because the gravel cover dissipates the impact energy. Some people claim that hail will cause damage to a BUR by driving the cover gravel into the felt or tar; except for the most extraordinary circumstances, **this is untrue.** Upon impact with a pebble, the pebble will transfer the impact energy to perhaps a dozen or more other pebbles, and so on. By the time the membrane layer is reached, the impact energy per pebble is trivial.

In most cases where it is claimed that hail has driven gravel into a BUR, the damage has actually occurred by people walking on the gravel and pushing it into the tar by their own weight (check for footprints!).

A gabled roof should have a noticeable difference in hail damage severity from one side of the roof to the other, due to the change in roof slope (impact angle).

Asphalt shingles have a hard coating of small stones to protect them from hail. As the shingles age, this coating is lost and the shingles lose their ability to withstand hail damage. When asphalt-shingled roofs are past their useful life, they may be damaged by even trivial hailstorms because they are essentially bare asphalt. Because asphalt will become brittle upon exposure to ultraviolet (U.V.) rays in sunlight, after a time even small impacts can cause damage. In such cases, it is recommended that shingles from roof areas not receiving hail be compared to "hail-damaged" shingles to evaluate the condition of the shingles prior to the hailstorm.

Wood shakes are not generally considered damaged by hail unless they have been cracked by impact. Damage in this case is defined as the loss of useful life or physical harm. Cracking occurs immediately upon impact and not in any delayed way. Hail-caused cracks are verifiable by the impact depression at the root of the crack. Hail blemishes are normal and do not harm the shake. In fact, they generally fade away.

Hail does not cause wood shakes to curl. This is caused by the grade of the shake, exposure to sunlight, average humidity, and age.

Old shakes that have been "softened" by bacterial attack will be easily damaged by hail, due to the low "e" value. Such shakes have not been damaged by hail; they were already damaged and in need of replacement before the hailstorm.

Built-up roofs without gravel cover can be damaged by hail due to the low "e" value of the tar. However, exposed BUR roofs are improper in the first place.

"Popped" blisters on built-up roofs are the result of a bad or worn-out roof, not of hail.

The new single-ply membranes cannot be damaged by hail unless severe deterioration has already occurred.

REFERENCES

1. American Insurance Association, *Hail Damage to Red Cedar Shingles*, 1975, 58 pp.
2. Sidney Greenfield, *Hail Resistance of Roofing Products*, Building Science Series 23, National Bureau of Standards, 1969.

CHAPTER 15

Drought and Its Effects

EFFECTS OF THE DROUGHT

Few areas of the United States have escaped the effects of the worst drought in over 50 years. At the end of September, 1988, the Kansas City area was still about 11 inches short of its average annual rainfall; this is a shortfall of more than 25 percent.

Besides well-publicized problems the drought has created in agriculture, some of its more subtle effects are yet to occur. In the months to come, as moisture levels return to normal, it is expected that settlement problems will greatly increase. These problems will occur not only in newer buildings, but also in some homes and structures that have been standing for decades without previous significant settling.

Some of the foundation problems, of which the root cause may be the drought, may be mistakenly blamed on nearby construction work, poor workmanship or blasting effects. Further, as a consequence of the expected rise in foundation problems, it is also anticipated that there will be a rise in consumer complaints about overly expensive or ill-advised repairs.

BASIC SOIL PROPERTIES

While the delineation of soil properties can be very technical, soil is basically composed of gas, moisture and solids. The amount of each component in a unit of soil and the various ratios between the components affect the ability of that unit of soil to bear a load.

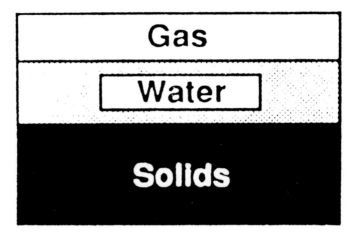

Figure 1. The basic components of soil.

From a practical standpoint, water is incompressible.   That is, under pressure, water does not reduce in volume. This is more or less also true of the soil particles. It is either the gas or the voids within the soil that give soil its "springiness". If the water is removed from a unit volume of soil, the volume of the soil will shrink.

THE ATTERBERG LIMITS

The Atterberg Limits is a system that defines how the bearing strength of fine-grained soil changes with water content.

The **liquid limit** is the largest amount of moisture a soil can hold at which the soil still has some resistance to shear. At the liquid limit, the soil has no cohesion and is essentially a strengthless mud.

The **plasticity limit** is the least amount of moisture content at which a soil is plastic, or pliant. The **shrinkage limit** is the least amount of moisture content needed to saturate the soil.

A typical soil with a void ratio of 0.8 (the volume of gas divided by the volume of solids) at the shrinkage limit will weigh about 122 lbs./cu.ft. The same typical soil with the same voids ratio with all the moisture removed will weigh about 95 lbs./cu.ft.

DIFFERENTIAL SETTLEMENT

As soil dries out, it normally dries out first near the surface. As dry conditions persist, the upper, dried-out soil will absorb moisture from lower levels of soil; then lower levels will also gradually dry out.

Under normal conditions, only a small upper layer of soil will become dry between rains. During a prolonged drought, the soil will lose significant moisture at much greater depths.

Thus, additions to a building which have a shallow foundation depth will be affected first and most adversely. Building portions which are supported by a deeper foundation will generally be affected last and least.

Consider a house with a basement foundation that has an attached front porch which is set on simple foundation blocks. During a prolonged drought, the porch will likely drop in relation to the house. This is because the house is supported by its basement walls, which are set in soil that is deep enough not to have lost much moisture. However, the front porch support blocks sit on soil near the surface that has dried out and shrunk. Thus, between the house and the porch one would find settlement cracks and problems.

The problem can be complicated by the underground drainage pattern. Even if the house has a foundation set at the same depth all around, a drought coupled with an uneven drainage pattern may cause one corner to lose moisture faster than the others, and drop.

WHEN IT RAINS AGAIN

When clay soil dries out, it shrinks. This usually results in the soil consolidating (settlement) or pulling away from areas. Sometimes, a significant gap will appear between the foundation walls and the edge of the soil where it has shrunk away from the wall.

Often, when the soil consolidates, it redistributes itself, filling in voids and "holes". In this way, the soil will still maintain contact with foundation walls. When the drought ends, the dry soil will expand as it reabsorbs moisture. This can further exacerbate differential settlement or heaving.

The soil resting against foundation walls will push laterally upon the walls. This lateral push may be greater than before, due to the drop of the soil. This may cause basement walls to be pushed inward. Typically, the walls will bow in the center and possibly crack at this point. Cracks will be wider at the top and narrower at the bottom.

This problem is often greatest with concrete-block walls. Block walls have little or no lateral strength. Thus, the soil expansion resulting from the drought ending will often crack and damage block walls, while poured basement walls could withstand the expansion.

In the first rain after a long drought, the rain may run down the "gap" that forms between the soil and the basement wall, causing the basement to leak, or to leak much more than it did before.

Case Study

A split-level house had a back yard higher than the front yard. The house was supported by a basement. The rear basement wall was fully below grade, but half of the basement wall was above grade. The front porch, which was a poured cement platform and stairs, was tied by reinforcing bars into the other half of the front basement wall, which was mostly below grade.

The house sustained the following damage:

1. The driveway was cracking.

2. The front porch was "drooping."

3. The front basement wall was bowed, outward at the top and inward at the center, at the point where the porch was attached.

4. The brick veneer on the front was separating from the wall.

What had occurred was that the front yard had dried out more than the back yard. The front porch had dropped, pulling on the top of the basement wall where it was attached, and was also pushing on the basement wall in the center. The porch was suspended like a diving board; it was bending around the connection point. The veneer was pulling away from the house because the basement walls above grade had dropped along the driveway.

Despite the fact that the house was 25 years old, most of the damage occurred within the last six months. The owner had initially blamed the damage on a construction project located about 200 feet away on the same side of the street. The construction project had begun in June, about the same time that the drought effects were becoming evident and severe.

OTHER EFFECTS

Besides the havoc that the drought and moisture cycle has caused and will continue to cause to building foundations, the relative soil movements and building movements have also caused additional stress on water pipes, gas pipes, and sewer lines.

In cases where the pipes are made of brittle materials, such as clay sewer lines, or in cases where the pipes have become delicate due to corrosion, the relative shifts in the ground may be enough to cause cracking and leakage.

In some areas, in-ground concrete swimming pools may also sustain this kind of settlement damage, especially if the pool was empty during the drought and the following rains. When empty, an in-ground swimming pool has minimal lateral wall strength. With the pool full, the water pressure will help equalize the pressure of the expanding soil.

CHAPTER 16

Lighting

SOME BASIC CONCEPTS

Light is simply one form of electromagnetic radiation, which also includes x-rays, radio and television signals, infrared radiation and cosmic rays. The visible light perceived by our eyes has a wavelength that ranges from 700 nanometers (the very red end of the spectrum) to about 400 nanometers (the violet end of the spectrum). A nanometer is one-billionth of a meter. Interestingly, visible light has a higher frequency than radar or radio waves, but a lower frequency than x-rays.

Light has many properties in common with an undulating wave. The distance from one crest to the next is called the wavelength. The number of wavelengths that pass a particular point in a specific period of time is called the frequency. If the wavelength is multiplied by the frequency, the transmission speed of the radiation is obtained.

Visible light, as well as other types of radiant energy, has a speed of about 300,000,000 meters per second, or 186,000 miles per second. The formula which relates wavelength, frequency and speed is as follows:

$$c = f \times l \tag{1}$$

where c = speed of light
f = frequency
l = wavelength

145

For example, a green light with a wavelength of 500 nanometers has a frequency of 600,000 billion hertz (or wavelengths per second).

When all visible wavelengths of light are combined in equal amounts, white light is produced. However, if the light is produced so that one, or a group of wavelengths, predominate, the light will be colored accordingly. Sunlight and incandescent lamps are usually considered sources of white light. Sodium lamps produce a yellowish light, and mercury lamps produce a blue-green color. Some manufacturers combine sodium and mercury lamps in a single fixture to produce a more "whitish" light.

When a beam of light is directed at a surface, it will usually "bounce" or reflect off the surface, especially if the latter is smooth and shiny. If the surface does not cause the light to disperse much, the reflection is called **specular**. When the reflected light is dispersed and has no specific reflected direction, the reflection is called **diffuse**. Rough surfaces generally cause the reflection to be diffuse.

The ratio of reflected light intensity to the incident light intensity is called **reflectance**, usually given as a percentage or a fraction (e.g., a material with a 40-percent reflectance will reflect away 40 percent of the light originally shined upon it).

Similarly, **absorption** is the amount of light that a surface will absorb, also given as a percentage or a fraction. Dark-colored materials with rough surface textures generally absorb more light than smooth, light-colored surfaces.

Transmittance refers to the amount of light that can pass through a material. Glass is the most notable example of a material with high transmittance. Of course, concrete has none.

The sum of reflectance, transmittance and absorption is always one. Materials which do not let light pass through them at all, like concrete, have a transmittance of zero. In such "solid" materials, the light is either reflected or absorbed, and the sum of the reflectance and absorption is then one:

$$1.0 = R + T + A \tag{2}$$

where R = fraction of light reflected
　　　T = fraction of light transmitted, and
　　　A = fraction of light absorbed.

Certain materials, such as glass, generally have high transmittance and low reflectance when the incident light beam is perpendicular to the surface. However, as the angle from perpendicularity increases, transmittance decreases and reflectance increases.

Because of its transmittance, glass allows natural light to come into an interior space during the day, and increases the amount of light available inside. At night, however, the glass allows light produced by light fixtures to "escape", and thus reduces the amount of light available.

MEASUREMENT OF LIGHT

In the English system of units, light intensity is measured in **footcandles**. One footcandle is the intensity of light from a standard candle as measured one foot away. A **lumen** is a measure of the total quantity of light; one lumen is defined as a footcandle of light falling on one square foot of area.   The relation between lumens and footcandles is as follows:

$$(1 \text{ footcandle}) \times (1 \text{ square foot}) = 1 \text{ lumen}$$

Luminance or **brightness** is the amount of light that is perceived to come from an item via reflection or transmission. In the English system, luminance is measured in **footlamberts**. If one footcandle of light falls on a perfectly reflecting surface (R = 1.0), then the luminance is 1 footlambert. If one footcandle of light falls on a surface which reflects only 50 percent of the incident light (R = 0.5), then the luminance is 0.5 footlamberts.

The luminance or perceived brightness of a light fixture that has a transmittance of 80 percent, a size of 2 ft by 4 ft, and four lamps that produce 4,000 lumens each, is as follows:

$$\frac{4 \times (4{,}000\,lumens) \times 0.80}{2ft \times 4ft} = 1{,}600 \; footlamberts$$

For practical purposes, light meters normally measure both the light intensity and luminance in footcandles.

One further important relationship is that the intensity of light from a single source drops off in proportion to the square of the distance. For example, if the light intensity is measured to be 40 footcandles at 10 feet from the light source, at 20 feet it would be 10 footcandles, and at 100 feet the light intensity would be 0.4 footcandles.

FACTORS IN AREA LIGHTING

In an interior area, the perceived light depends on several factors:

1. The direct light from the lamp fixture.

2. The reflected light from the walls, ceiling, and floor;

3. Any contribution of natural light through windows, and

4. The angle and direction from which the light is seen.

Because the ceiling can be lighter (more reflective) than the floor, because the light source can be more speculative than diffuse, and because a person can look directly into the lamp or away into a shadow, the perceived light intensity can vary greatly even at the same spot in a given room.

For example, a parking garage with unpainted concrete ceilings, floors, and walls might have a light reading of 5 footcandles at a particular spot and direction. With everything the same except that a 10' by 10' patch above the light fixture is painted white, the light reading might measure 7 footcandles.

Light from a lamp fixture will also vary due to its design. Most types of fixtures have a published chart, called a candlepower distribution chart, that indicates what the light intensity will be at various angles from the lamp. Such charts are used by architects and engineers to plan the illumination for an area. Additionally, each type of lamp has an expected life in hours which is useful in estimating the frequency of lamp failure for maintenance purposes.

LIGHTING'S ROLE IN ACCIDENTS

Some accidents occur because there is too little light to see properly. Slip-and-fall accidents and industrial accidents are often blamed upon inadequate lighting. However, some accidents occur because there is too much light, or it is in a color that is not easily discernible. For example, when an approaching car has very bright high beams, or its lights are colored to resemble the tail lights of a truck, a driver may not be able to properly respond, and an accident may occur.

For offices and schools, the Illuminating Engineering Society (IES) has established comprehensive recommendations for lighting. These recommendations include the amount of reflectance in walls and ceilings, the amount of light perceived at various angles (sometimes referred to as the "scissors curve"), and the amount of light directed to task areas.

Other general lighting recommendations exist, that indicate what the average lighting should be for various types of work areas. The following general table indicates the range of brightness (luminance) required for various tasks.

Table 1. General Task Luminance

Task Type	Luminance Needed
Casual	0 - 20 footlamberts
Normal	20 - 50 footlamberts
Detailed	50 - 100 footlamberts
Very Detailed	100 - 400 footlamberts

The usual lighting recommendations given in footcandles refer to horizontal working plane, and are referred to as the general lighting design criteria. The following table lists a few typical values.

Table 2. Typical Recommended Lighting Levels

Type of Area	Lighting Level
Locker Room	30 footcandles
Parking Garage	5 footcandles
Office Area	70 footcandles
School Auditorium	30 footcandles
Barber Shop	100 footcandles
Bathroom	10 footcandles
Drafting Room	100 footcandles
Kitchen	70 footcandles

ACCIDENT CHECKLIST

When inappropriate lighting may have caused or contributed to an accident, the following items should be checked and noted:

1. The exact position of the person when the accident occurred.

2. The time and date of the accident.

3. The condition and type of lights at the time (how many were on, how many were burnt out, etc.).

4. The activity of the person at the time of the accident.

5. The usual recommended level of lighting for the type of area and activity.

6. The visual ability of the person (can he/she see well enough, does he/she require glasses and wear them, etc.).

As long as the above is known and no changes have been made to the scene, measurements of the lighting can be made and compared to the recommended levels. It is very important to do this right away, because many lighting-related accidents are the result of poor maintenance; too many lights have burnt out. After an accident, it is an easy matter to replace the burned-out lamps and restore the proper lighting level.

CHAPTER 17

Slip-and-Fall Accidents

"Slip and fall" is the catch phrase generally used to describe fall-down accidents that occur on publicly accessible walking surfaces. Typically, the accident victim has fallen down and sustained an injury, and claims that the surface in question was too slick to walk on safely.

In making a claim about the floor slickness, the victim normally wishes to transfer liability for the accident to the owner or caretaker of the floor. Monetary claims often include current and projected medical expenses, and current and projected lost earnings. Additional sums for pain and suffering or loss of consortium may also be claimed.

STATISTICAL BACKGROUND

The following table summarizes the number and type of accidents resulting in death, which occurred in the U.S. in 1988; note the total for falls.

From this table, it can be seen that deaths from falls in the home slightly outnumber deaths from falls that occur outside the house. The total number of deaths by falls outside the home is similar to the number of deaths by drowning, by poisoning, or by fires and burns.

Table 1. U.S. Accident Deaths in 1988

Type of Accident	Number of Deaths
Motor Vehicles	48,000
All falls (all kinds)	12,000
Falls in the home (all kinds)	6,500
Drownings	5,000
Fires/Burns	5,000
Food Poisoning/ Ingestion of Objects	3,600
Firearms	1,400
Poisonings (solid or liquid)	5,300
Poisoning by gas	1,000

As might be expected, deaths from falls are more frequent in the elderly than in younger persons. Older people may have infirmities which can be exacerbated by a fall, and often attempt actions they are no longer capable of doing safely. This expectation is borne out by the statistics in Table 2.

Interestingly, the number of men who died from falls was 53 percent of the total, indicating that total deaths from falls are about evenly split between men and women. As a comparison, men account for 71 percent of deaths from motor vehicle causes, 83 percent of deaths by drowning, and 86 percent of deaths from firearms.

The age group of 65-plus constitutes about 12 percent of the population of the U.S.; however, it accounts for 73 percent of deaths from falls. Of this same group, 41 percent are men and 59 percent are women.

Table 2. U.S. Deaths from Falls by Age - 1988

Age Group	Deaths from Falls
0 - 5 years	117
5 - 14 years	55
15 - 24 years	399
24 - 44 years	1,035
45 - 64 years	1,519
65 - 74 years	1,554
75 + years	6,765

Only 14 percent of deaths from falls occurred to people under 45 years of age; 86 percent occurred to persons over 45. From this it is obvious that age is an important contributing factor in fall deaths irrespective of other circumstances.

COEFFICIENT OF FRICTION

The "slickness" of a floor is normally measured by its **coefficient of friction** with another material. Friction is defined as the resistance to force caused by contact with another material. For example, when a person has to push a large box along a floor, the person must push hard enough to overcome the resistance of frictional force caused by contact of the box with the floor.

The coefficient of friction is defined as the ratio of the force needed to overcome the friction to the force perpendicular to the contact surface. For example, if the box weighs 200 pounds, and it takes 75 pounds of force directed parallel to the floor to "break it loose" and get it going, then the COF between the box and the floor is

COF = frictional force / weight

= 75 lbs./ 200 lbs.
= 0.375

There are two common types of friction: dynamic and static. **Dynamic friction** is associated with the resistance to force when an object is already moving along; **static friction** is associated with the force needed to "break loose" an object and get it going. Static friction is usually numerically greater than dynamic friction. Some readers know from experience that it is easier to push a car to keep it rolling than it is to start it rolling. With respect to slip-and-fall cases, it is the static coefficient of friction that is normally of concern.

It is important to note that the coefficient of friction is not a rating for a single material, but is a value that exists between two contacting materials. If one of the materials is changed, the value will change. For example, a running-type shoe on a tile floor may have a COF of 0.55, while a lady's dress high-heel shoe may have a COF of 0.15 on the very same floor.

The following table lists some representative static COF's for some common material combinations.

Table 3. Some Common Static Coefficients of Friction

Material Combination	C O F
Car tire on dry concrete	0.70
Hard steel on hard steel	0.78
Steel on steel, greased	0.05
Wood on wood	0.55
Leather on wood	0.60
Car tire on snow-packed surface	0.15
Wood runners on ice	0.03

MEASURING COF

For a particular shoe and floor combination, the technique for measuring the COF is simple. The shoe in question is loaded with a known weight of sand. The shoe is then pulled forward or backward, using a spring scale to measure the amount of force needed for it to begin to move. Due to the difference in shoe sole geometry, there can be differences between the COF measured while pulling the shoe forward as opposed to pulling it backward, or even sideways. The ratio of the weight noted on the spring scale to the weight of the shoe and sand is the static COF.

For example, a 4 oz. shoe is filled with 15 pounds of sand. Pulling forward, the shoe required 6 pounds of pull force to break it loose; its forward static COF is then:

$$COF = 6 \text{ lbs}/15.25 \text{ lbs} = 0.39$$

A second method of measuring the COF involves using a standardized "shoe" on a sample of the floor in a James Machine. The method is described in detail in ASTM D-2047-82, *"Standard Test Method for Static Coefficient of Friction of Polish-Coated Floor Surfaces as Measured by the James Machine".* While the above is a laboratory method, there are also a number of portable James machines which can test surfaces on site. However, the portable models generally do not give the same test accuracy.

The method in question uses a standardly prepared and sized piece of leather to act as the shoe. The leather sample is weight-loaded and pulled over a floor sample panel in four different directions, 90 degrees to one another. Three prepared panels of flooring material are required for each floor test. The reported result is the average of all twelve static COF readings.

This method is normally used by manufacturers to test different wax and polish floor coatings. In general, a wax or polish coating that provides a COF of 0.5 or more can be labelled slip-resistant or slip-retardant. However, the standardized leather piece used in the test procedure normally gives much higher COF values than do actual shoes. A COF determined by this method should not be confused with the actual COF of the shoes and floor in question; they can be quite different.

STANDARDS FOR FLOORS

Standards for floors vis-a-vis friction are very vague; essentially there are no specific standards. In a monograph published by the National Bureau of Standards, *An Overview of Floor Slip-Resistance Research with Annotated Bibliography* (NBS Technical Note 895), various COF ranges are cited and discussed. For example, a government deck covering standard requires a static COF of 0.30 against sanded leather. The 0.50 value noted above was also cited in the NBS report.   The problem with these values is that they are based on testing by a James-type machine on standardized leather or rubber samples; they do not represent actual shoe-to-floor values.

Page 24 of the NBS report reviews a study where a value of 0.23 is cited as the mean lower value of static COF between floor and shoe to prevent slipping at the start and stop of a person's step. The start and stop of a step is generally where a person is most vulnerable to slip. Notably, since the value cited is a mean or average value, some values above and below it were also noted in the study at which slip did not occur, and walking took place safely.

The *Uniform Federal Accessibility Standards* (49 FR 31528) states that "ground and floor surfaces along accessible routes and in accessible rooms and spaces including floors, walks, ramps, stairs and curb ramps, shall be stable, firm, slip-resistant, and shall comply with 4.5 (Ground and Floor Spaces)." It could be argued that since ASTM D-2047 defines slip-resistant as a floor surface with a static COF of at least 0.5, as measured by the D-2047 method, then the slip-resistant reference could be construed to provide an actual COF number for floor surfaces.

However, the definition of "slip-resistant" given in ASTM D-2047 refers to qualifying a floor polish product to use the term in commercial advertisements (see rule 5, "The use of terms slip-retardant, slip-resistant, or terms of similar import" in the Trade Practice Rules of the Floor Wax and Floor Polish Industry, issued by the Federal Trade Commission, March 17, 1953). The linkage of the two specifications, therefore, is tenuous.

The problem with setting floor standards is two-fold: not all floors should have the same COF, and not all shoes are alike. If a dance floor had the same COF as a boat deck, it is likely that no one could dance on it.   While people also slip on carpets, carpets are rarely blamed for being too slippery. A carpet with a COF of 0.60 would be like walking on Velcro and would be very difficult to clean.

Interestingly, while victims of slip-and-fall accidents often sue the owner or caretaker for the floor being too slippery, they rarely sue the manufacturer of the shoes. Yet, the shoe is just as important in determining the COF as the floor surface itself.

CHOICE OF SHOE

Most of us already know that new dress shoes are more prone to slide than old dress shoes or tennis shoes. Of course, the decision to wear such slick-bottomed dress shoes instead of tennis shoes or flats lies solely with the wearer. Actual tests have found that a floor surface with a COF of 0.60 against running shoes can have a COF of 0.17 against new, high-heel ladies' dress pumps. Shoe manufacturers apparently do not have to adhere to any COF standards with respect to sole slickness. Because of this, it is difficult to select or prepare a floor that will have a high COF for all types of shoes.

In the case of high-heel dress shoes, a slip-and-fall accident may not have actually involved a slip, but may have involved buckling. It takes practice for a woman to walk around in high-heel shoes without toppling or wobbling. Because of this skill factor, girls, elderly women, and persons with weak ankles have difficulty in maintaining balance. Twisted ankles and falls are not uncommon, irrespective of the floor surface.

A narrow or spiked heel can also fit into small crannies, and can hang up on small edges that would cause no problems with ordinary shoes. In the *Uniform Federal Accessibility Standards*, carpet edges of 1/2 inch in height, and grating holes which are 1/2 inch wide and any length, are allowed in public-access areas. These dimensions are larger than the heel base of many common types of narrow-heeled shoes.

FLOOR AND SHOE COATINGS

Normally, a wet floor is slicker than a dry one. However, in some cases a slightly wet floor may actually increase traction. Damp leather, for example, tends to grip floor surfaces better than dry leather. This type of anomaly is why on-site testing of the actual shoe on the actual surface is important in investigating an accident.

Other nonliquid materials can also affect floor friction. Fine dust or sand on either the floor or the shoe can adversely affect traction. Tennis shoes on tile normally have good traction, but tennis shoes with a light coating of dust on the soles will more easily slide on the same tile floor. For this reason, gymnasts commonly have to wipe the soles of their tennis shoes between events to ensure proper traction.

Since high-traction shoes are often more prone to pick up a dust film, it may be important to know where a person has been walking prior to a slip-and-fall accident.

The choice of wax or polish can increase or decrease the slickness of a floor. As noted, some polishes are rated as being slip-resistant. Unfortunately, a high-friction floor will generally be harder to clean than a slick one. Floor types that may normally have good traction may become slippery if the wrong wax is applied. Thus, the type of wax or polish applied can be another important factor in such cases. As before, actual testing of the shoe and floor after the accident will provide relevant data in this respect.

CHAPTER 18

Water Pipe Freeze-ups

A COMMON SCENARIO

The following is an all-too-common sequence involving water leaks: A family leaves for a holiday and is away from their house for several days. Upon their return, they find that the carpets, walls, ceilings and the baby grand piano are all water soaked. Damages are extensive.

The leak is traced to a water pipe located in the upstairs bathroom. Because of the cold weather while they were away, the claim is made that the leak was caused by a frozen pipe that ruptured.

CORRELATION IS NOT CAUSATION

Leaks in water pipes can occur for many reasons, many of which do not involve freezing. Failures which do not involve freezing can, of course, happen both in summer and winter. Thus, the mere fact that a pipe leak occurred in a cold spell does not automatically prove that freezing was the culprit.

For example, in Topeka, Kansas, the months of November, December, January, February and March all have (on the average) minimum daily temperatures below freezing. Also, the months of December, January and February have average daily temperatures below freezing. Thus, on the average, five months of the year in Topeka have freezing temperatures. Assuming that a non-freeze-related water-pipe failure can occur at any time, then it has a 42 percent chance of taking place during a month that normally has frost.

ICE VERSUS WATER

In general, when materials become cold they contract, and when they are warmed up they expand.   Water, however, takes some liberties with this rule; from 32 degrees Fahrenheit (freezing) to 39.2 degrees, water actually contracts slightly.   From 39.2 degrees F to higher temperatures, water expands more or less in accordance with the increase in temperature.

Table 1. Water Density versus Temperature

Temperature	Density (lb/cu.ft)
32.2	62.4183
35.6	62.4246
39.2	62.4266
42.8	62.4246
64.4	62.3407
86.0	62.1568
100.4	61.9893

For the most part, water is considered to be incompressible. Literally this means that it cannot be compressed when subjected to external pressure; however, this is not strictly true, as Table 2 shows.

For the range of pressures encountered in typical residential plumbing, say less than 150 psi, water can be assumed to be incompressible to simplify calculations, with little effect on accuracy.

Table 2. Compressibility of Water

Temperature	Pressure, psia	Density lb/cu.ft
32° F	14.7	62.4183
	7,350	62.8943
	14,700	65.2502
	29,400	67.6768
68° F	14.7	62.3186
	7,350	63.6662
	14,700	64.8906
	29,400	67.0300
122° F	14.7	61.5294
	7,350	62.9534
	14,700	64.1372
	29,400	66.2052

For most common materials, the solid state is heavier than the liquid state. For example, a chunk of steel will not float in a pool of molten steel, because its density is higher. Once again, though, water is the exception to the rule: when water solidifies into ice, its density is significantly less than its liquid density, and thus it floats. Once frozen, ice itself will expand and contract with temperature, the same as most materials.

(As an aside, if ice and water behaved like most other materials, ice would sink, and the ocean depths would be filled with ice. The earth would be a much different and very peculiar place.)

At 32 degrees F, liquid water has a density of 62.4 lbs./cu.ft. However, at 32 degrees, solid water - ice - has a density of about 57.2 lb./cu.ft. This is an increase in volume of 9.3 percent for the same weight at the same temperature.

This sets the stage for the basic problem: when temperature drops, most materials used for pipes contract slightly, and water will also contract, for the most part, along with the pipes, until it is cold enough to form ice. But when ice forms, it will expand to a volume 9.3 percent larger than it occupied before, at the same temperature. This has the effect of dramatically increasing the internal pressure in the pipe.

PIPE STRESSES

In a long, thin-walled pipe closed off at both ends, the **hoop stress** (the stress which would cause the diameter of the pipe to expand or pull apart) can be calculated by the following formula:

$$S_h = \frac{(P \times D)}{(2t)} \tag{1}$$

where S_h = hoop stress,
     P   = internal pressure in pipe,
     D   = nominal thickness of the pipe, and
     t   = wall thickness of the pipe.

Thus, if the water pressure were 100 lbs/sq.in. in a 4-inch diameter pipe with a wall thickness of 0.125 inch, the hoop stress would be 1,600 lbs/sq.inch.

However, the pressure in the pipe also causes a second stress, called the **axial stress**. This stress is perpendicular to the hoop stress, and is directed along the length of the pipe. Axial stress causes the length of the pipe to increase. The axial stress of a pipe is calculated by the following formula:

$$S_a = \frac{(P \times D)}{4t} \tag{2}$$

where S_a = axial stress,
and the other symbols remain the same.

Thus, if the same piece of pipe is considered as before, the axial stress caused by the same internal pressure of 100 lb/sq.in. is half as much as the hoop stress, or 800 lb/sq.inch.

If all the water in the pipe were to become solid ice, the volume of the pipe would have to increase by 9.3 percent, or the pipe would rupture. By the above formulas, it can be seen that if the pipe ruptures, it will rupture first along a line that follows the length of the pipe. This is because the hoop stress is twice as much as the axial stress, and the hoop stress acts to pull the pipe apart across its diameter. It will be the hoop stress that first exceeds the material strength of the pipe. Thus, a freeze-over in a pipe will generally cause the pipe to split along its length, assuming that the ends are at least as strong as the walls.

If the pipe is made of a ductile material, such as mild steel or plastic, its walls may have thinned down prior to rupture. The pipe material would stretch outward to accommodate the increased volume. However, this thinning effect can be masked by corrosion or wear. In brittle materials, such as cast iron, no thinning will occur.

FRACTURING IN STEPS

When water freezes in a pipe, it does so relatively slowly. First, the water at the wall of the pipe freezes. It then thickens and advances toward the center of the pipe. The process is like a slow choking of the pipe.

However, it may not be necessary for the pipe to fully freeze over for the pipe to crack or split open. When enough ice has formed on the pipe wall interior to increase its internal volume beyond the pipe's ability to expand without rupture, a small split will occur. The initial split will be just long enough to relieve the pent-up stresses which have developed up to that point. When more ice is formed, the split will lengthen accordingly.

In this fashion, the split will lengthen in steps. Thus, a pipe split caused by a freeze will exhibit a number of start-stop or sawtooth signature patterns in the fracture. These are usually observable under low magnification.

A pipe which has split because of a water hammer, or other sudden overpressures, will fracture all at once (i.e., brittle fracture mode). There will not be a start-stop pattern in the fracture, because there was only one sudden overpressure.

A fracture pattern which shows evidence of fatigue, or low-cycle overload, indicates that there are problems in the mounting of the piping. A fracture which tends to follow a spiral indicates that the pipe was in a bind and was being twisted. This is also unrelated to a freeze-up, and has to do with the way the piping was installed.

PIPE ENDS

When the pipe ends in a fitting that is less strong than the pipe wall, such as a compression fitting, the fitting will be pushed off the pipe, instead of the pipe being split open along its length. In order for this to occur, the amount of internal pressure needed to dislodge the fitting must be less than that needed to split the pipe wall.

As the pressure fitting is pushed off, it too will leave a start-stop pattern along its contact areas, which will generally be observable under low magnification. A pressure fitting that is blown off by a simple water hammer or overpressure will generally show a simple, one-event abrasion pattern along its contact areas, which can be observed.

If the threads in a threaded pipe are cut too deep at the ends, it is possible that the end cap or joint may break off during a freeze-up. In such cases, the fracture will generally begin at the deepest thread (where the remaining wall is thinnest) and will follow the thread most of the way around. However, for this to happen, the thread must be cut deeper than half the thickness of the pipe.

INSULATION

In order for insulation to keep a pipe from freezing, it is necessary that there be some heat present. If the water in the pipe is not moving, and the pipe is not heated in some way to keep its temperature above freezing, insulation wrap by itself will not help. A cold pipe is still a cold pipe even if it is well insulated. There must be a heat source somewhere for the insulation to be of any use. Further, when there is a heat source, the insulation must be placed between the pipe and the cold, not between the pipe and the heat, as sometimes happens.

A common mistake in new homes occurs when the insulation is put on the wrong side of the piping, such as piping near or in an exterior wall. The insulation should be on the side of the pipe that faces the cold, not the side that faces the heat.

FREEZE-UP PRIORITY

Pipes that freeze up are generally the pipes in exterior walls, crawl spaces, open attics, or within finished areas where air infiltration is high. When such pipes freeze, normally there is an order or pattern to the way they freeze. Look at the big picture first.

For example, if the outside temperature is 20 degrees F, the pipe to the toilet may freeze up first. As the temperature drops even more, other pipes in the house may freeze up in some kind of orderly sequence.

If a reported frozen pipe has "frozen" out of sequence, this indicates that some other problem may be at work, or that something has changed.

All things remaining the same, a pipe freeze-up will normally occur on the coldest day to date in the winter season. For example, let us suppose that a reported freeze-up occurred in February when the temperature dropped to 15 degrees F for two days. If, however, in January the temperature had dropped to 5 degrees F for three days, then the claimed freeze-up would be out of sequence. It should have more logically occurred in January during the more severe cold spell.

When such an out-of-sequence freeze-up is reported, either the conditions around the pipe have significantly changed, or the problem is not due to a freeze-up.

A common weather pattern associated with frozen pipe/water damage problems is when the coldest freeze to date occurs, followed by a thaw. The pipe may split open during the freeze-up, but no water damage will occur at that time because the frozen water cannot flow. When the thaw comes, the ice melts and the water damages occur. Thus, an extended freeze can temporarily hide frozen pipe damages which may have already occurred.

NOAA (The National Oceanic and Atmospheric Administration) compiles weather data for most metropolitan areas in the United States. High and low temperatures for each day are recorded, as well as the temperature at three-hour intervals. This type of data is very useful in comparing the freeze/thaw patterns of the local weather to the particulars of a water damage/frozen pipe report. This data is available through most libraries, or can be ordered directly from NOAA in Asheville, North Carolina.

CHAPTER 19

Principles of Machinery Guarding

HISTORICAL BACKGROUND

One of the outgrowths of the European Renaissance was the scientific revolution. Science and technology, which had more or less been at a standstill during the Middle Ages, made great gains once the restraints of religious dogmatism were loosened. Galileo, Copernicus, Kepler, Newton, Hooke, Descartes, and Boyle are just a few of the outstanding persons of that era who provided the scientific underpinnings upon which modern technology is based. Of course, once basic scientific principles were understood and disseminated, it did not take long for them to be applied for economic gain.

The Industrial Revolution began in the early 18th century as an outgrowth of the scientific revolution. In essence, the Industrial Revolution was the application of science to industry. The meager power generated by beasts and men was replaced by the more abundant and forceful power from machinery. Likewise, the products made by individual labor were superseded by those made by scientifically designed machinery.

Each new invention, in turn, allowed the creation of even more productive machines. The Newcomen steam pump, invented in 1712, allowed more coal per man to be mined. The Watt steam engine, patented in 1769 and powered by coal, became a substitute for power previously provided by water wheels, men and beasts. Because of the steam engine, factories could now be located away from the banks of rivers, near the sources of raw materials or markets, thus reducing transportation costs. The inventions of the flying shuttle (1736) and the spinning jenny (1764) eliminated much of the need for manual labor in textile production, thus automating production.

With this new-found technology, it became possible to build large textile mills, employing first tens, then hundreds, and later thousands of people. Each of these developments profoundly affected history. In fact, the military pre-eminence of Great Britain during the 19th century can be directly attributed to the pre-eminence of her textile manufacturing prowess.

However, the Industrial Revolution also brought with it new hazards. The many gears, cogs, belts, and turning shafts comprising powered machinery could move much faster than a person. Improvements in production efficiency were often accompanied by a commensurate speeding up of the machinery. An operator could not depend on his reflexes and agility to keep out of harm's way. As the machines got bigger, they became more powerful. Machinery had become not only faster than the operators, it was also stronger. Once snagged by machinery, an operator was often not strong enough to free himself before being injured.

By the middle of the 19th century, the problem of **industrial accidents** due to inadvertent contact with unguarded machinery was enormous. Employers were virtually unrestrained by any safety regulations, legal or social responsibilities. Workers were subordinate in importance to the machinery they were operating. Injuries were so common that it was possible to guess a person's trade by the nature of their scars and lost limbs. In some industries, children were preferentially hired; their small size and nimbleness better suited the available space around the machinery.

In response to this and other problems, labor organizations came to be developed. The first labor union in the United States, the Knights of Labor, was organized in Philadelphia in 1869. Perhaps predictably, its membership came initially from the garment manufacturing industry, which was the most automated. Later, it included workers from many industrial sectors. By 1886, it had 700,000 members.

Various other labor organizations were formed concurrently, some using violence to achieve their aims. The terrorist Molly Maguires, an organization of Irish coal miners from the coal fields around Scranton, Pennsylvania, murdered various mining officials and law officers who protected the mining interests, in an attempt to gain better working conditions. In 1877, eleven of their leaders were hanged for those offenses, and the organization was effectively broken up.

From 1869 to the present time, the history of industrial safety is a complex mixture of politics, economics, law, social movements, engineering and sometimes violence. It is a subject worthy of study by itself. For our purposes, however, it is enough to appreciate that current standards for machinery safety are not simply the product of some conscientious technical committee; they are, rather, the accumulated result of historical confrontations and compromises in various industries.

OSHA

Current U.S. machine-guarding regulations stem from the Williams-Steiger Occupational Safety and Health Act, passed by the U.S. Congress in 1970. This act established the **Occupational Safety and Health Administration (OSHA)**, and required OSHA to codify comprehensive industrial and construction safety standards by the end of 1972. Since a number of excellent standards already existed at the time, OSHA incorporated many of them into the OSHA code. With respect to machinery guarding, much of the OSHA code was incorporated from standards previously published by the **American National Standards Institute (ANSI)** and by the **American Society of Mechanical Engineers (ASME)**.

The Williams-Steiger Act is a landmark in U.S. industrial safety, because it established nationally uniform industrial safety standards. Before 1972, safety regulations were a patchwork varying by state, union jurisdiction, industry, and civil-law precedent. Importantly, the Act also established an orderly method for the modification or deletion of current standards, and for the adoption of additional standards. The method takes into consideration new research findings, and employs an open hearing review process to consider industrial, labor and public concerns.

In general, the machinery guarding regulations are contained in OSHA regulations 1910.211-1910.222, and 1910.263(c), revised as of

July 1, 1990. Additionally, OSHA has published a booklet, *Concepts and Techniques of Machine Safeguarding*, (OSHA 3067, 1980) which more simply explains and illustrates the various ways of guarding machinery.

Often related to guarding matters, OSHA also requires certain signs and markings to designate hazards. These are contained in OSHA regulations 1910.144 and 1910.145.

MACHINE GUARDING BASICS

In general, any machine component or operation that could cause injury must be effectively guarded. Most commonly, mechanical hazards are associated with the following three situations:

•   Point of Operation: The point in the machinery where cutting, shearing, shaping, material removal, or forming operations are performed.

•   Power Transmission Equipment: The machinery components that transmit power through the machinery. This includes sheaves, belts, connecting rods, cams, couplings, chains, gears, cogs, flywheels, spindles, sprockets and shafts.

•   Other Moving Parts: This includes the moving parts of the machinery not directly associated with the first two categories, such as feed mechanisms, augers, flyball governors, moving fixtures, control mechanisms etc.

An easy rule to consider with respect to guarding is this: would you want to put your own face in it? If not, you had better provide protective guarding.

Direct Barrier

A machine guard can take many forms. The most common is the **direct barrier**. A barrier simply prevents contact being made with the dangerous item while it is operating. A common example is the chain guard on a bicycle; it prevents inadvertent contact of a person's leg or clothing with the bicycle chain and sprockets. Without the chain guard, clothing or skin could become entangled at the pinch point between the chain and sprocket, or snagged by the irregularities of the moving chain links.

For a barrier to be an effective guard, a person should not be able to easily remove it, and it should be durable. In some industries where pay is based on piecework, an operator may be tempted to remove a guard in order to save time, and increase his pay. Of course such removal may substantially increase the risk of injury. Thus, an easily removable guard constitutes an "invitation" to circumvent the guard. Similarly, a guard which easily breaks apart or wears out is not considered effective. The guard must stand up to normal wear and tear, including bumps and impacts from tooling which might be associated with routine work in that area.

To protect the operator from his own inadvertent action or those of others, guards which might be removed occasionally by the operator for legitimate reasons can be equipped with trip switches or interlock devices, which will disable the operation of the machinery when the guard is removed. Thus a person who is engaged in resetting some mechanism and is obscured from sight will not accidentally be injured by another person who unknowingly attempts to operate the machinery.

A barrier guard should also cause no new hazard, and should allow the operator to work unimpeded. Guards made of sheet metal with sharp edges may create a new laceration hazard where there was none before. Guards which get in the way of work or aggravate a worker often are circumvented; workers will connive to get rid of them if possible.

Guarding by use of direct barriers is often considered the preferred method of guarding. This is because no special training or actions of the operator are needed, and the hazard is eliminated at the source. However, some situations do not easily lend themselves to this method and require alternative guarding techniques.

Protective Clothing

The basic premise of guarding is to prevent exposure to hazardous things in the workplace. Besides the use of a direct barrier, this can also be done in some instances with protective equipment or clothing.

Protective equipment or clothing is especially useful where there is a danger of chemical splash, high-intensity noise or light, noxious vapors, or similar hazards which cannot be fully eliminated by good practices or design.

Such items as face shields, ear plugs, helmets, protective cloaks or coats, respirators, goggles, and steel-toed shoes are in this category. Whenever possible, however, the root source of the hazard should be eliminated or reduced by design and engineering.

Safety Devices

In instances where the machine's hazardous equipment cannot be effectively safeguarded by a fixed barrier, various electromechanical safety devices can be utilized to prevent contact.

For example, photoelectric sensors can be used to determine if a person's hands or body is in a danger zone. While they are in this zone, the machine is disabled. Such a device is often used in power brake presses, where the duration of action is very short, but the force of the forming action is sufficient to severely maim or sever body parts.

Pull-back devices are often used on machines with stroking actions. A pull-back device allows an operator to manipulate a work piece at the point of operation when the punch or shear blade is in the up position. When the machine begins its downstroke, cables attached to the operator's wrists automatically ensure that his hands are pulled clear of the work area.

Alternatively, a two-handed trip can often accomplish the same effect. In such a system, the operator must have both hands (and sometimes feet) simultaneously pressing independent switches before the machine will operate. This ensures that the operator's parts are out of harm's way.

Trip wires and trip bars are devices that disable a machine when depressed or activated. These devices are used when operators have to work near equipment pinch points or moving machinery, and there is a risk of falling into the equipment.

While not strictly guards, other safety devices that prevent inadvertent injury by machinery should not be overlooked. Such devices include: safety blocks and chocks, lock-outs on power disconnect boxes, lock-outs on pneumatic and hydraulic valves, push sticks and blocks for feeding materials, etc. (See OSHA 1910.147.)

Guarding by Location

A machine can also be safeguarded by location; that is, it can be so placed that a person cannot normally get to it, or to its hazardous equipment. Access to hazardous machinery can be denied by fence enclosures, or by fixed wall barricades. Safety can be even further assured by placing a safety switch in the door or gate which will stop the machinery when its enclosure is opened.

EMPLOYER'S RESPONSIBILITIES

In a nutshell, it is an employer's responsibility to ensure that all equipment used in his shop meets current OSHA safety standards.

Older machines built before current OSHA standards were enacted must be brought up to current standards. Similarly, imported equipment that does not meet current OSHA standards must be modified to do so. Unlike an older car or building, manufacturing equipment is not "grandfathered" for safety purposes.

Employers must regularly inspect equipment as per OSHA regulations, and maintain records of such inspections. Worn-out guards and safety devices must be promptly repaired.

Consider the following: A belt guard was also equipped with an interlock switch so that when the guard was removed, the belt and pulleys would not operate. The switch wore out. To expedite production, a "jumper" wire was run to bypass the switch to keep the machine going. The switch was then forgotten. Two months later, an operator wished to adjust the belt tightness. He turned off the machine and removed the guard. While he was making adjustments, another person mistakenly turned on the machine, and the operator was injured. In such a situation, it is likely that the employer will be held accountable for the accident, rather than the employee who mistakenly turned on the machine without first checking.

Employers must also be watchful for employees who circumvent guards and safety equipment. If it can be shown that the employer knew about or failed to reasonably correct such equipment abuses, the employer can be held accountable for any ensuing accidents.

Work rules are not a substitute for guarding. An employer cannot simply promulgate a work rule that employees must keep their hands and limbs away from dangerous equipment, or that employees should work safely. An employer cannot depend on notions of "common sense" by his employees to avoid accidents and injuries.

In general, an employer cannot assume a passive role with respect to shop safety. He must become informed of standards and requirements, and be actively involved. In the legal sense, since he is inviting people to his workplace, they have a reasonable expectation that they are being provided with reasonably safe equipment.

MANUFACTURER'S RESPONSIBILITIES

The manufacturer of machinery has, of course, the responsibility to design and construct his equipment so that it is in compliance with current OSHA standards. Also, the equipment should be in keeping with current industry standards and practices. A commonly used legal expression is "state of the art design", which is usually taken to mean that the item under consideration is as good and as bad as everybody else's product currently being sold.

But, there are persons (mostly plaintiff attorneys) who will argue that this is still not enough. They will argue that the various published standards are just a minimum; that a machine is not deemed safe simply because it meets all published standards and regulations.

There is, of course, some merit to this argument. As new machines are designed and built with innovative features, faster feed rates, greater power, etc., new hazards appear which must be dealt with. For example, the original OSHA safety standards published in 1972 had not yet fully addressed the use of high-powered lasers in manufacturing, which were just being introduced at the time. It would have been specious to argue that a patently unsafe laser sheet-metal cutter met all industry standards when there weren't any.

In this context then, it is incumbent on the manufacturer to at least recognize the "spirit" or underlying intent of safety standards, and to make reasonable efforts to ensure that a machine is safe and properly guarded, even if standards or regulations don't yet exist.

The designer of machinery would do well to consider that even the best operators make really "bonehead" mistakes when they are tired, bored, or simply asleep. Given enough operating hours, a mistake that would be considered "one in a million" might occur several times. Is it a good design if a person could lose a hand because of simple (or even stupid) operating mistake?

CHAPTER 20

The Engineer as Expert Witness

INTRODUCTION

To the lay public and even to many engineers, engineering is often considered an objective science. This is perhaps fostered by the quantitative problem solving methods used by engineers in design work. People hire engineers to tell them exact answers to their questions. For example, how many cubic meters of earth must be dug out, how much steel is required, or what size bearing is needed?

Undergraduate training of engineers in the classroom also emphasizes exact problem-solving techniques. Students spend many hours calculating the correct answer, learning the correct fact, or applying the correct theory to answer their homework problems. When there is doubt, often the correct answer will be found listed at the back of the textbook.

Engineers like to refer to their discipline as a "hard" science; one that provides a "hard" or exact solution to a problem. This is on contrast to disciplines like psychology, sociology or economics, which engineers would consider "soft" sciences because of their inability to supply exact or specific answers. While an engineer can calculate when a beam will break due to excessive load stress, a sociologist is unable to calculate the date when a community will riot because of analogous social stress.

Because of this traditional prejudice, some engineers are wholly taken aback when they present their case findings and conclusions in a courtroom for the first time. Their well-reasoned, scientifically sound investigation of an accident or catastrophic event may be pronounced unsound or fallacious, and may even be dismissed out of hand. In fact, their qualifications, which may impress other engineers, will be belittled; their experience, which may be considerable, will be minimized, and their character and professionalism, no matter how impeccable, will be questioned. It is not unknown for the experience to be so unpleasant to some that they never again undertake a forensic assignment.

In a courtroom, extremely well qualified and distinguished professional engineers may testify on behalf of one side of an issue. They may radically disagree with another set of equally well qualified and distinguished engineers who may testify on behalf of the other side of the issue. Some people might presume that the spectacle of strong disagreement among practitioners of such a "hard" science indicates that one side or the other has been bought off, is incompetent, or is just outright lying.

While the engineering profession is certainly not immune from the same dishonesty that plagues other professions, the basis for disagreement is often not due to corruption or malfeasance. Rather, it is a highly visible demonstration of the subjective aspects of engineering. Nowhere else is the subjectivity of engineering so naked as in a courtroom. To some engineers and laypersons, it is embarrassing to discover, perhaps for the first time, that engineering does indeed share some of the same uncertainties as, say, sociology.

ADVERSARIAL NATURE OF OUR LEGAL SYSTEM

The most fundamental point that must be understood about the U.S. legal system is that it is **adversarial**. The attorney for each party involved is obliged to argue as persuasively as possible for the benefit of his client.

It is not the duty of the attorney to judge his client; that is the prerogative of the judge and jury. However, it is the attorney's duty to be his client's advocate. In one sense, the attorney **is** his client; the attorney is supposed to do for his client what the client would do for himself had he the same training and experience. When all attorneys in a dispute present their cases as well as possible, the judge and jury can make the most informed decision possible.

Because of this adversarial role, no attorney will allow another party to present evidence hurtful to his client's case without challenging and probing its validity. If the conclusions of an engineer expert witness cause his client to be out ten million dollars, it is a sure bet that the attorney will not let those conclusions stand unchallenged!

The above principle should be well considered by the forensic engineer in all aspects of an assignment. It is unreasonable for a forensic engineer to expect that his testimony or report will not be assailed and questioned, no matter how distinguished his credentials.

INCOMPLETE INFORMATION

Honest professional engineers can and do disagree. As was discussed in a previous chapter, forensic engineering differs from other scientific endeavors which use the classical scientific method. This is because the event being analyzed is not controlled and usually cannot be repeated for additional study.

There is no opportunity to apply, in the regular way, the classical scientific methodology of observation, hypothesis, experimentation and theory. In fact, since the event is unexpected or accidental, there may have been no opportunity to obtain in-progress data, as would occur in a controlled experiment.

In the reconstruction of an accident or catastrophic event, there is no singling out of factors as in a controlled experiment. In a controlled experiment, the relative influence of various factors can be repeatedly measured and quantified. In a one-time catastrophic event, many factors act at the same time or in unknown sequences. Different effects can produce the same results. Thus, when only the result is known, the causative effect or effects may not be unique.

The reconstruction of an accident or a catastrophic event is like curve fitting: when the data points are few, many possible curves can be hypothesized to fit the data. When the data points are plentiful, the variation in curve types to fit them is reduced.

Thus, there are inherent systematic reasons why two otherwise honest and competent engineers may disagree about how and why a catastrophic even occurred.

SAMPLE SCHEDULE OF EVENTS IN A LITIGATED DISPUTE

A simple dispute might occur as follows: A warehouse fire takes place, and a forensic engineer is engaged by the insurer of the warehouse, insurer "A", to determine the cause and origin of the fire. At the same time, the area fire marshall might also be investigating the fire, as might another insurer, "B", who insured the contents of the warehouse on behalf of a third party.

After a time, the engineer for "A" completes his investigation and concludes that the fire was caused by faulty wiring that had recently been installed. Insurer "A" will consider this information and the conclusions of the fire marshal. Possibly, insurer "A" may also have information developed by insurer "B". Insurer "A" may even hire a third engineer to review the findings of the other two engineers and of the fire marshal. Insurer "A" may also take statements from witnesses, involved parties, etc., and then confer with legal counsel.

When insurer "A" has gathered sufficient information that leads him to believe that the electrical contractor is at fault, he may send a demand letter to the contractor's insurer - insurer "C". This letter will state that insurer "A" believes the electrical contractor was at fault for the fire, and how much money is required to remedy the losses.

A point should be noted here. In most cases, insurer "A" will have already paid out compensation to the warehouse owner under the condition of his policy. Insurer "A" is seeking a monetary settlement from insurer "C" in order to put the blame where it belongs, and mitigate his own company's losses.

If the case put forward by insurer "A" is compelling, insurer "C" may cave in and pay up. If the case is less than compelling, but is still convincing, the two companies may negotiate to split the loss in some way. However, if insurer "C" believes insurer "A" has no case, or at least no case that can be won in a courtroom, insurer "C" may deny the claim.

Often, the dispute is settled at this point. Experienced adjusters and attorneys know what makes a winnable case, and what doesn't. However, if no satisfactory resolution of the claim is found, insurer "A" will file a suit. Insurer "B" may join in the suit with insurer "A" as co-plaintiffs and to pool resources; this helps defray the court preparation costs for both.

After suit is filed, a period of time called "discovery" occurs. This is when all parties to the dispute disclose what evidence they have to prove their respective positions.

One of the purposes of discovery is to encourage settlement before trial by making everyone, so to speak, show their cards. However, since disclosure is done in response to specific inquiry, there is still some intrigue that can take place if the right questions have not been asked.

When the inquiries are written, they are called **interrogatories**. Oral inquiries take place in a conference called a **deposition**. Usually, when a person is summoned to a deposition, he is **subpoenaed**. When the subpoena is a "duces tecum", the person has to bring all files, records and physical evidence relevant to the case in his possession or control.

A deposition is like a mini-trial. Many of the rules of procedure and decorum of trial are observed. Present at a deposition will be the attorneys, a court reporter, and, in this example, the engineer. The judge may be present if he wishes, but rarely exercises this prerogative.

The court reporter makes a permanent record of everything said, and swears in the engineer, now called the **deponent**. All attorneys in the dispute are allowed to ask the engineer about his conclusions, opinions, qualifications, observations, experiences, etc., relevant to the dispute. However, usually the attorney for "the other side" does most of the questioning. In case the engineer cannot appear at court, the record of his deposition can be read. His deposition can also be used to impeach him at the trial, if his trial testimony is significantly different.

After all the important witnesses and parties have been deposed, all parties again have time to assess the strengths of their positions and to reach a settlement prior to trial. Often, a close look at what the "other side" has to offer in evidence and witnesses is enough to close the differences between the parties. However, if both parties are still recalcitrant, the case goes to trial.

THE TRIAL

A basic fact of civil trials is that the evidence requirements for a finding of guilt are not as stringent as in criminal cases. Unlike a criminal case, where there has to be unanimity among jurors, a civil case may require only a two-thirds majority. In a criminal case the judge may charge the jury to find a guilty verdict only if there is no doubt about the evidence, but a civil jury can return a guilty finding based only upon a preponderance of evidence.

The plaintiff in the case must show why the defendant is guilty or the wrong or loss. On the other hand, the defendant does not have to prove his innocence; he has only to prove that the plaintiff is in error in blaming him for the loss. If the attorney for the defendant can cast enough doubt on the veracity of the plaintiff's account of the cause of loss, that may be enough for the defendant to be found innocent of the wrong, or tort.

Civil court trials rarely happen promptly after an accident; a typical trial may take place three or four years later. For this reason, the engineer must document his findings and information well enough to comprehensively brush up on the facts before the trial. A person cannot rely on memory alone.

In civil disputes, a trial is somewhat akin to a school play: before the trial, the judge and the attorneys have already agreed on the "program" to be followed at the trial. Each attorney has given the judge a summary of his witnesses, what they will say, and what evidence will be produced. If the discovery process was done properly, there will be few surprises for either side at the trial.

It is not necessary to go into all the particulars of trial procedure here; only the portion relevant to the engineer as expert witness will be discussed. However, it is important to know that the case is developed by the attorneys asking a series of questions of witnesses and other involved parties. The attorneys themselves do not directly present testimony or evidence.

When the engineer is called to the witness stand, he is sworn and asked to sit down. The first person to ask questions will be "his" attorney; this is called **direct examination**. The attorney will ask questions in a planned sequence to bring out the qualifications, experience, observations, findings, conclusions and opinions of the engineer relevant to the dispute.

The next person to ask the engineer questions will be the attorney for the "other side"; this is called **cross-examination**. During "cross", the attorney will try to tear down or dispute what the first attorney established.   Usually he will focus on points he believes are the weakest in the engineer's testimony. He may find fault with the engineer's qualifications or experience.   He may dispute facts or principles on which the engineer has relied. He may present information about the case previously unknown to the engineer, or pose hypothetical questions.   He will try to get the engineer to agree to opinions or facts that are supportive to the "other side's" position, or to the "other side's" engineer expert.

When the attorney for the "other side" finishes his cross-examination, the engineer's "own" attorney may again ask questions; this is called **re-direct**. He may do this if he thinks some point brought up by the other side needs explaining, or to clarify some information that looks bad for his side. He does not have to engage in "re-direct"; if he does not, the questioning is over and the engineer is excused.

However, if he does exercise his right of re-direct, the attorney for the "other side"may also exercise his right of **re-cross examination**. In both re-direct and re-cross, the attorneys are supposed to discuss only items that have been already brought up in direct and cross; no new lines of questioning are supposed to be developed.

If needed, a third, fourth or fifth round of re-directs and re-crosses can occur. The questioning ends when both sides have had the same number of opportunities to ask questions, and the engineer's attorney stops. The engineer is then excused.

However, sometimes the engineer may be brought back to testify as a **rebuttal witness**. This would happen if the "other side's" expert testified to something that was incorrect, unscientific, or in error. The engineer would then be called back to the stand to discuss only those points in dispute, to rebut the other expert's testimony.

STRATEGIES USED TO DISCREDIT ENGINEERS

Attorneys tend to use the same tried-and-true methods to discredit expert witnesses that oppose their cases. Here are a few of the more popular techniques.

Qualifications: During questioning, the attorney for the other side may attempt to show that the engineer's education, experience or licensing are not sufficient to warrant his designation as an expert in this matter.

> Remedy: Undertake only those assignments for which you are qualified. Think ahead of time about why you are qualified and be prepared to prove it.

Money: The attorney will usually ask by whom the engineer is being paid, and how much per hour. The purpose is to infer that the engineer is a "hired gun" whose opinion is bought. In a poor county or town, the hourly rate of engineering fees can appear very high.

> Remedy: Be matter-of-fact about the business and answer straightforwardly. Most jurors know that people are paid for their time. It would be more suspicious for someone to do it free.

Previous Court Appearances: The attorney will want the engineer to appear to be a "court whore" who goes from trial to trial and has no other income. He may use the question, "How long have you been a professional witness?" or similar. He may produce advertisements from the engineer's company that purport to solicit court testimonial work.

> Remedy: Don't be a "court whore". A certain number of court appearances is inevitable, but they should be adjunct to your main work. Even lawyers are not in the courtroom all the time.

Deposition Impeachment: This is one of the most popular devices. The attorney will attempt to show that your answers at trial are different than at the deposition. He will set up the engineer by asking a question in court, and then asking the engineer to recall what he said (possibly years ago) in deposition. Sometimes the questions will be slightly different, but will sound the same.

> Remedy: Tell the truth all the time. You have the right to ask the attorney to read the question in context from the deposition to the court, so that any variances in the two questions can be made plain. Beware of compound questions, where you may answer yes to the latter part, but the attorney interprets your yes for the former part. When you are asked a compound question, answer the two parts separately. Beware of questions with an underlying assumption. (Famous example: Do you still beat your wife?)

Comparison of Qualifications: This occurs when the "other side" has a PhD or a distinguished expert. The attorney will first ask if the engineer thinks that a PhD is better than whatever credentials the engineer has, and then asks how the engineer could disagree with someone the engineer admits is more qualified. (Note: this type of questioning is generally inadmissible, but often it gets in anyway in one way or another.)

> Remedy: Be sure of your facts and information. If you cite a fact, state the nature of its legitimacy. For example, "as stated in the *Handbook of Chemistry and Physics*" or "as shown in photograph 3..." . One devious way to get around this is to research what the PhD has published and find your own citation in there.

Plausible Alternatives: One way the attorney can cast doubt on the engineer's findings is to invent a plausible alternative that has not been discussed by the engineer in his report, and against which there is no apparent disputing evidence in the record. As an example, if the engineer concluded that an explosion resulted from a leaking gas pipe, the attorney might ask if the engineer checked to determine if sewer gas might have been the culprit, and will point out how close a sewer drain was to the explosion epicenter. Remember, the defendant does not have to prove his innocence; he only has to prove that the plaintiff's case **might** be wrong.

Remedy: During the course of initial investigation, it is important to note things that did not cause the accident. In fact, it is just as important to document what did **not** cause the accident, as what did.   Prior to deposition or trial, the plausible alternatives should be anticipated and evidence cited to preclude them from serious consideration.

Accumulation of Small Errors: The attorney may find a number of small typos, minor errors of fact, etc., in the engineer's report or testimony. He may then argue that the work product is sloppy and error prone, and will then rhetorically ask what other errors or mistakes are imbedded in the engineer's evidence and testimony. This technique is especially effective if the attorney gets the engineer to admit to one important error but then affirm that there are no others. Of course, a prepared attorney will have several more such errors identified and will wait with baited breath to show the engineer those errors, dragging the process out to really make the engineer look careless.

Remedy: Turn out professional work. This is not an undergraduate paper where partial credit is given; this is the real world where people's money and reputations hang in balance. When mistakes are noted, admit to them in a matter-of-fact way. But it is far better not to have your nose rubbed in them in front of a jury and possibly even some reporters.

Deposition Impeachment - Part II: Some very resourceful attorneys will ask the engineer in deposition if he has testified before in similar case, but for the opposite side. He will then obtain the deposition of the engineer or the court transcripts to see if the engineer has given contradictory testimony. He will then present this information in court (usually as an effective surprise) and impeach the engineer's testimony. It will usually make the engineer appear to have sold his testimony to please his current client.

Remedy: Tell the truth all the time. Don't be sucked into being an advocate for the client; that is the attorney's role. Your role is to provide honest and competent engineering information and analysis. If you have furnished opinions in the past that appear to contradict opinions in this case, be thoroughly prepared to explain the differences. You have the right to ask the attorney to cite the full context from which he is quoting, if needed.

Apparent Ignorance: One of the favorite tactics of an attorney to show that the engineer is unqualified to give testimony is for the attorney to learn some of the terminology, facts and methods common to the field in question, and then castigate the engineer when he does not have them also memorized. For example, in a fire case, the attorney may learn the ignition points of several materials that were in the fire. During cross examination he may ask the engineer expert about these memorized facts to show that the engineer doesn't know even these simple facts. Of course, his expert will have these committed to memory and will recite them nonchalantly to impress the jury that knowing them is no big thing, and that everyone who is anyone in the field knows this stuff.

> Remedy: Do your homework. It doesn't hurt you to know this stuff. Take some reference materials to the trial. When you are asked something trivial but not committed to memory, ask in a matter-of-fact way if you can consult a reference book to provide the exact answer the attorney seeks. Take your time in looking it up. If the attorney does this several times, he will look like a jerk for wasting the jury's time.
>
> Similarly, if an equation has been used in the proceedings, the attorney may ask the engineer what effect changing some of its variables will have on the result. Be prepared to give that information. Don't look like a klutz fumbling with numbers or asking if anyone in the room has a calculator. You can be sure their expert will have the answers tucked away somewhere so that he can impress the jury. You can avoid this problem entirely by making a sensitivity analysis of the variables in your report. This makes your report appear even more thorough if the attorney has forgotten and asks.

Background Check: An attorney can easily discredit an expert witness if the witness has previously been convicted of felonies. Wife beating, child abuse, drug addiction, white-collar crimes, military-service screw-ups etc. will warm the cockles of the opposing attorney's heart.

> Remedy: Be a good person. If you have a checkered past, you had better prepared a heart-wrenching story of how you turned your life around. Also, if you have been successfully sued for deficient design work, your credibility will stink. Don't fool yourself into thinking no one will check; they will, and they will present it at the most opportune possible time to discredit your testimony. If you have problems in your past, be sure to discuss this well beforehand with the attorney who engaged your services; perhaps their effect can be minimized.

Provoking the Engineer: Sometimes an attorney will attempt to provoke or exasperate the engineer by his method of questioning. The idea is to make the engineer appear confused, angry or unprofessional. The attorney may develop a rhythm of rapid-fire questions all of which are answered "yes" except for the zinger which is slipped in and has a "no" answer. Or, the attorney may ask questions in an insulting manner.

> Remedy: Stay calm. Always take a reasonable time in responding to questions. Break up the attorney's rhythm. You do not have to follow his pace; set your own. Wait until the question is fully asked before responding. Remember that it is the attorney's job to do this. If the questions seem stupid, be patient and respond to them as best possible. After a while the jury will become irritated and the attorney will stop. He also risks currying the jury's sympathy for you if he drags this technique out.

Technical Arguments: The best way to bore a jury to pieces is to engage the experts in an arcane technical argument. The attorney may do this hoping that the jury may miss the more important points of the engineer's testimony while they are sleeping or counting holes in the ceiling.

Remedy: Present you findings and conclusions in simple terms. Use everyday things with which most people are familiar to demonstrate principles or facts. For example, when discussing the melting effects of an electrical short, draw a parallel with arc welding. A good approach is to couch testimony in the same language you would use to teach an uninformed person about the subject. Remember, the more information you bring up, the more the attorney for the other side has to work with in cross-examination.

SOME NOTES ABOUT LAWYERS AND ENGINEERS

An engineer cannot accept a "cut" of the winnings or a bonus for a favorable outcome. He can only be paid for his time and expenses. If it is found out that he has accepted an expert testimony assignment on some sort of contingency fee basis, it is grounds for having his professional engineer's (P.E.) license suspended or revoked. The premise of this policy is that, if he has a stake in the outcome, an engineer cannot be relied upon to give honest testimony in court.

Attorneys, on the other hand, can and do accept cases on a contingency basis. It is not uncommon for an attorney to accept an assignment on the promise of 30% - 40% of the "take" plus expenses, if the suit is successful. This is done so that poor people who have meritorious cases can afford legal representation.

This situation can create friction between the attorney and the engineer. First, the attorney may try to delay paying the engineer's bill until after the court case is over. This is a version of "when I get paid, you will get paid" and may be a de facto type of contingency arrangement. It is best to agree beforehand on a schedule of payments from the attorney, for services rendered. Follow the rule, "Would it sound bad in court if the other side brought it up?"

Secondly, since the lawyer is the advocate for the case, and may have a great opportunity for financial improvement, he may pressure the engineer to manufacture some theory to better position his client. If the engineer caves in to this temptation, he is actually doing the attorney a disservice.

The engineer does his job best when he informs the attorney of all aspects of the case, both good and bad. The "other side" may also have the benefit of an excellent engineer who will certainly point out the bad stuff in court. If the attorney is not properly informed of the bad stuff, he cannot properly prepare his case for court. There is nothing like being made a fool of in court to make an attorney appreciate the lack of backbone in his expert engineer.

Earlier, it was said that a court trial is like a school play. In this respect the attorney is the stage manager. It is his case to win or lose, not the engineer's. The engineer is simply one of the many actors who have parts to play. The forensic engineer does his job best when he is honest and thorough. He should avoid getting caught up in the advocacy of the case. During trial, when asked about things detrimental to his client's case, the engineer should answer succinctly with a professional demeanor.   If he hesitates or weasels around the question, the jury will think that the engineer is not objective in his testimony, and cannot be trusted.

CHAPTER 21

Reporting
the Results of
Investigative
Engineering

An accident has occurred for which an insurance claim has been made. Because of the technical and legal aspects of the claim, the claims adjusted has engaged the services of a professional engineer. The engineer visits the accident scene, photographs the important details, perhaps makes some small tests, and checks some references. He is now ready to put together his findings in a written report.

At this point, the engineer has to decide upon a format for his report. He could merely write what he has done in a simple narrative, starting from when he received the telephone call from the claims adjuster until the last item in the investigation was completed. Alternatively, he could prepare the report like a college paper, replete with technical jargon, equations, graphs and footnotes to impress his client with his professional prowess.

WHO READS THE REPORT?

Before he decides, let us consider who reads his report and for what it may be used. The first person who reads the report is the claims adjuster. He wants to know what happened and, more importantly, whether this is a claim which his company might have to pay. The adjuster, then, needs to be told what happened in terms that he can relate to the policy language.

The next person to read the report might be the attorney for the insurance company. He wants to know if there is something about the claim for which he can subrogate. He needs to know if there were design shortcomings, material defects, workmanship deficiencies, omitted warning signs, or other similar problems. He needs these items clearly described so that he can relate them to the law.

If the lawyer for the insurance company decides that there is potential for subrogation, the report will also be read by the notorious "other side," their attorneys and their technical experts from some famous college. They will want to know upon what facts and observations the engineer relied, which regulations and standards he consulted, and what scientific principles or methodologies were used to reach the conclusions about the cause of the loss. They, of course, will want to pick the report apart. They will challenge each and every facet of the report in an attempt to prove that it is a worthless sham.

The next person to deal with the report may be, surprisingly, the engineer who authored it. Several years after the original investigation, the engineer may be called upon to testify in deposition or court about his findings, methodology, and analytical process. He may have to rely on his own report to remind himself of the particulars.

Eventually, the report may end up in a closed jury room where jurors from all types of backgrounds will read it and personally interpret what is says, and does not say. The author will not be there to explain any ambiguities; it will have to stand on its own.

These are the things that a good investigative or forensic engineering report must do. Often, such a report for so many potential readers, must be written in the context of a $300 or $500 job budget and within a short time from when the assignment is given. The engineer also has the burden of completing the report in short order so that the adjuster does not bump up against the state's "good faith" laws.

In order to accomplish all of these aims, the author's firm uses a report format based on the classical style of argument, as used in the Roman Senate almost two thousand years ago. It follows this general outline:

1.   Purpose: A succinct statement of the assignment.   For example, to determine the cause and origin of a fire.

2.   Brief Background: Some general information as to what happened and to whom so that the reader understands what is being discussed.

       Section 2 lists the facts in the case. It says who did the work, what his qualifications are to do the work, when he did the work, how it was documented, who witnessed it being done (x'ed out), and what was observed. Factual observations are referenced to the appended photographs. ("If you don't believe me, here's my photographic proof.") Item G reports information provided by the insured. It is important to cite such informational sources; in three years no one may remember who said what.

3.   Finding and Observations: A list of all factual findings and observations made related to this matter. No opinions or analysis are included in this sections. Just the facts, ma'am.

       Section 3 in the example tells and explains the opinions of the investigating engineer. The first paragraph summarizes the section. The second paragraph explains why the house fuses did not stop the shorting. The third paragraph describes the reasoning as to the cause of the shorting and fire. Possible violations of the electrical code are cited, and it is noted that the wiring which shorted was newly installed. This paragraph may have obvious interest for the lawyers.

4.   Analysis: This is the section where the engineer gets to "show off". The facts are analyzed and the engineer's conclusions are explained to the reader. Highly technical calculations or extensive data are listed in full in an appendix, but the salient points are summarized and reported here for the reader's consideration.

Section 4 states the three conclusions reached by the engineer: where the fire started, why it started, and what is to blame.

5.   Conclusions: In just a few sentences, the findings are summarized. For example, "the fire was caused by a new toaster that had a bad switch." This section should be written like the last page of a murder mystery: "The butler did it! (gasp)". Actually, if the report has been written cohesively up to this point, the conclusions should be already obvious to the reader.

Section "Remarks" in the sample report was added to attest to the fact that the investigating engineer did not alter any items at the scene. Often, the "other side" will allege that the investigating engineer altered the evidence during the investigation, rendering it inadmissible for evidence later. ("Your Honor, the evidence in its present condition does not faithfully represent its condition just after the fire and was deliberately altered by the engineer.")

6.   Miscellaneous: Sometimes an additional section is required to clean up side matters. For example, "Oh, by the way, your building may fall down and kill you next week"; or "We have the incriminating evidence at our place", or "Put a guard on that machine before any more poodles are sucked in."

7.   Attachments: photographs, tables, calculations etc.

In some cases, it may be necessary to alter the evidence in order to assess it; perhaps open a box lid which crumbles or remove some soot to see the nameplate. In such cases, the exact nature of the alteration should be noted, and it would be a good idea to photograph it both before and after the alteration.

8.   The signature identifies the author of the report and the reviewer (if any). In the author's firm, the person who writes the report has done the most work to accomplish the job, but in other firms this is not always the case. In some firms, notes are given to a technical writer to write the report, and the

writer may not have done any direct work on the assignment. Also, in some firms, the P.E. who signs the report may not have actually taken part in the work. He may be just "signing off" on the work of someone else, who may not be named in the report.

This is an important point to note if there are legal proceedings. The P.E. who signs the report is likely to be the person who stands up to explain the report to a courtroom. It is worth knowing whether this person actually had a real part in the work.

Additionally, the report is not a professional engineering report, unless it is signed by a duly licensed professional engineer, signified by a "P.E." after the name. This is an important legal distinction; only licensed professional engineers are allowed to sell engineering services to the public. Some states accord licensed professional engineers a "Friend of the Court" status.

Some companies purport to provide technical services, consulting services, investigative services, or scientific consulting services, and their reports may be signed by persons with various initials or credentials after their names. These designations have varying degrees of legal status or legitimacy. Thus, it is important to know the professional status of the person who signs the report. A report signed by a person without the needed credentials to testify in court might as well not be written.

It is worth remembering that a "good" case can be lost or overlooked because of a poorly written investigative report.

SAMPLE REPORT

PURPOSE

To determine the nature and cause of residence fire.

FINDINGS AND OBSERVATIONS

Dr. Randall Noon, a licensed professional engineer, traveled to xxxxxxxxxxxx and examined the xxxxxxxxxxxxxxx residence on April 11, 1991. xxxxxxxxxx was also present during Mr. Noon's visit. Photographs were taken to document observations and are appended to this report. The photographs are numerically indexed to explanatory captions.

The following observations were made:

A. The origin of the fire was a fuse box in the rear wall of the house in the kitchen. A classic "V" burn pattern was observed around the fuse box (Photograph 7). No other such burn pattern was observed in the house.

B. The fuse box used older-type screw socket fuses (Photograph 1). All of the fuses were fire or heat damaged; several had the fuse lens glass cracked out.

C. The wooden framing around the fuse box was deeply charred at the top (Photographs 1-5). Charring was not as severe along the sides of the frame.

D. Beaded wiring was found inside the fuse box on the incoming feeder conductors (Photograph 4). The wiring had arced over and cut through the top of the steel box.

E. Beaded wiring was also noted in conductors at the top of the box on the outside (Photographs 6 and 8).

F.   The feeder line from the meter box was noted to be beaded, and a char pattern followed the feeder line back to the meter box (Photographs 9-12). The meter and meter box did not appear damaged.

G.   xxxxxxxxxxxxxxx indicated that the meter box and weatherhead had been moved recently to the rear of the house. The local utility required the move.

ANALYSIS

The fire originated within the fuse box in the kitchen. The fire was caused by shorting of the incoming feeder conductors to the fuse box. The conductors were shorted to the top of the metal fuse box, cutting the steel box at that location. Shorting then followed the feeder back to the meter box.

Since the shorting occurred on the incoming side of the fuse box, the fuses could not open the circuit and protect the house. In order to arrest the short, the fuses on the service pole had to open.

The shorting occurred in new wiring which had been recently installed from the meter box to the fuse box. The location of the initial shorting, just on the inside of the incoming knockout opening, indicates that the feeder conductors may have been "skinned" during installation, or that they may have become cut on the edge of the knockout hole in the top of the box. No anti-abrasion ring or grommet was observed at the top knockout of the fuse box (note Photograph 3) as is required by the National Electric Code (NEC).

CONCLUSIONS

Based upon an examination of the premises, it is the professional opinion of Dressler Consulting Engineers, Incorporated (DCEI), that:

A.   The fire originated in the fuse box in the kitchen;

B.   The cause of the fire was shorting between the incoming feeder conductor and the top of the metal fuse box.

C.   The fuse box was not noted to have an anti-abrasion grommet or ring at the top of the fuse box as is required by the NEC. Such an item would normally have been installed in the fuse box at the time of the installation of the feeder conductor. The item prevents damage to the insulation of the feeder conductor by abrasion of the wire against the edge of the knockout hole.

REMARKS

The fuse box was left intact in the wall of the residence. No alterations or destructive testing was done by DCEI.

RECOMMENDATIONS

The fuse box and framing should be removed from the wall and stored when repairs are made to the house if any legal actions are contemplated.

DRESSLER CONSULTING ENGINEERS, INCORPORATED

Randall Noon, Ph.D., P.E.   Donald G. Dressler, P.E.
Director   President
Engineering Services

DGD/RKN/jdh

Attachment: Photographs

INDEX